Introduction

Collins Gem Geography is one of a series of illustrated
dictionaries of the key terms and concepts used in the
most important school subjects. For this edition the text
has been updated and colour is now used throughout
the book. With its alphabetical arrangement, the book is
designed for quick reference to explain the meaning of
words used in the subject and so provides a companion
both to course work and during revision.

Bold words in an entry identify key terms which are
explained in greater detail in entries of their own;
important terms that do not have separate entries are
shown in *italic* and are explained in the entry in which
they occur.

Other titles in the *Basic Facts* series include:
Gem *Biology*
Gem *Chemistry*
Gem *Computers*
Gem *Business Studies*
Gem *Mathematics*
Gem *Modern History*
Gem *Physics*
Gem *Science*
Gem *Technology*

GEOGRAPHY
BASIC FACTS

HarperCollins*Publishers*

HarperCollins*Publishers*
Westerhill Road, Bishopbriggs, Glasgow G64 2QT

First published 1983
Fifth edition 2002

Reprint 10 9 8 7 6 5 4 3
© HarperCollins*Publishers* 2002

ISBN 0 00 713803 2

Designed and typeset by Book Creation Services Limited

Printed in Italy by Amadeus S.p.A.

A

abrasion   The wearing away of the landscape by rivers, **glaciers**, the sea or wind, caused by the **load** of debris that they carry. *See also* **corrasion**.

abrasion platform   *See* **wave-cut platform**.

accessibility   A measure of the ease and efficiency with which a location can be reached. Central locations are highly accessible; peripheral ones are not.

acid rain   Rain that contains a high concentration of pollutants, notably sulphur and nitrogen oxides. These pollutants are produced from factories, power stations burning **fossil fuels** and car exhausts. Once in the **atmosphere**, the sulphur and nitrogen oxides combine with moisture to give sulphuric and nitric acids which fall as corrosive rain. The results of acid rain have been devastating for many lakes, forests and buildings in Scandinavia and Germany, due to **prevailing winds** in this part of the northern hemisphere blowing from a southwesterly direction, bringing acid rain formed in the industrial centres of western Europe. *See* **pH**.

administrative region   An area in which organizations carry out administrative functions. For example, the regions of local health authorities and water companies, and commercial sales regions.

adult literacy rate   A percentage measure showing the proportion of an adult population that can read. It is one of the measures used to assess the level of development of a country.

aerial photograph   A photograph taken from above the ground. There are two types of aerial photograph – a vertical photograph (or 'bird's eye view') and an oblique photograph where the camera is held at an angle. Aerial photographs are often taken from aircrafts and provide useful information for map-making and surveys. *Compare* **satellite image**.

afforestation   The conversion of open land to forest; especially, in Britain, the planting of coniferous trees in upland areas for commercial gain. Coniferous afforestation is currently a controversial issue in the Flow Country of northern Scotland, where it has destroyed large parts of this unique blanket bog area which is the **habitat** of many rare plants, birds and animals. **Conservation** efforts are now being made to protect what is left of this **ecosystem**.

    In areas of the world at risk from **desertification**, the planting of suitable trees helps to slow down **soil erosion**, and thus the expansion of the desert, by both binding the soil and providing **rainfall interception**. The trees also shelter the land, and any crops, from the force of the wind, and produce humus to increase the fertility of the soil (*see* **horizon**). *Compare* **deforestation**.

age and sex structure   The classification of the elements of a national or regional population according to sex and age groups. Age and sex structure determine the shape of a **population pyramid**.

agglomerate   A mass of coarse rock fragments or blocks of lava produced during a volcanic eruption.

agglomeration   The tendency for firms with related products to locate close to each other in order to reduce transport costs and other overheads: for example, the motor manufacture and component industries, and the oil refining and petrochemical industries.

agribusiness   Modern **intensive farming** which uses machinery and **artificial fertilizers** to increase **yield** and output. Thus agriculture resembles an industrial process in which the general running and managing of the farm could parallel that of large-scale industry. Many farms in East Anglia could be called agribusinesses. There, small family farms have been bought up by large food-processing companies located near the farms, where the farm produce is canned, processed, frozen, etc.

A 'branch' of agribusiness is factory farming, i.e. the intensive rearing of animals such as pigs, cattle and poultry. Many of these animals spend their lives penned in special rearing units where food, water, light and temperature are carefully controlled. Thus, factory farming can be seen as a production line, the sole aim being to increase output and profit. Many people

disagree with this form of farming, but it is argued
that less intensive methods mean higher prices for
the consumer.

agriculture   Human management of the **environment**
to produce food. The numerous forms of agriculture fall
into three groups: **commercial agriculture, subsistence
agriculture** and **peasant agriculture**. *See also*
agribusiness.

aid   The provision of finance, personnel and
equipment for furthering economic development and
improving standards of living in the **Third World**. Most
aid is organized by international institutions (e.g. the
United Nations), by charities (e.g. Oxfam) (*see* **non-
governmental organizations** (NGOs); or by national
governments. Aid to a country from the international
institutions is called *multilateral aid*. Aid from one
country to another is called *bilateral aid*.

 Aid can be *short-term* to meet an immediate crisis,
long-term to further development, or *tied*, linking aid
expenditure directly back to the donating country.

alluvial fan   A cone of **sediment** deposited at an
abrupt change of slope; for example, where a post-
glacial stream meets the flat floor of a **U-shaped valley**.
Alluvial fans are also common in arid regions where
streams flowing off **escarpments** may periodically carry
large **loads** of sediment during **flash floods**.

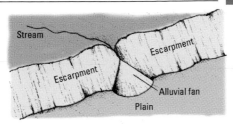

alluvial fan

alluvium Material deposited by a river in its middle and lower course. Alluvium comprises **silt**, sand and coarser debris eroded from the river's upper course and transported downstream. Alluvium is deposited in a graded sequence: coarsest first (heaviest) and finest last (lightest). Regular floods in the lower course create extensive layers of alluvium which can build up to a considerable depth on the **flood plain**. Some of the world's most fertile lands are found on alluvial flood plains.

alp A gentle slope above the steep sides of a glaciated valley, often used for summer grazing. (*See figure overleaf.*) See also **transhumance**.

amenity resources Those **resources** which provide an opportunity for recreation and leisure pursuits. The **national parks**, forests and the coastline are good

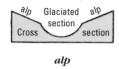

alp

examples of these, as are areas of parkland and open space in cities.

antarctic circle *See* **latitude.**

anemometer An instrument for measuring the velocity of the wind. An anemometer should be fixed on a post at least 5 m above ground level. The wind blows the cups around and the speed is read off the dial in km/hr (or knots).

anemometer

anthracite A hard form of **coal** with a high carbon content and few impurities.

anticline An arch in folded **strata**; the opposite of **syncline**. *See* **fold.**

anticyclone An area of high atmospheric pressure with light winds, clear skies and settled **weather**. In summer, anticyclones are associated with warm and sunny conditions; in winter, they bring frost and fog as well as sunshine.

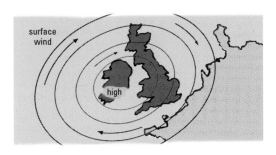

surface wind

high

anticyclone

appropriate technology Techniques and equipment which are appropriate to the immediate needs of a **developing country**. For example, ox-drawn planters and reapers, made of local materials, may be more useful in developing **agriculture** in such countries than tractors and combines, which are expensive to run, difficult to maintain in remote regions, and may cause unemployment. Appropriate technology stresses the need for low cost, straightforward and **labour-intensive**

projects, as opposed to the **capital-intensive, high-technology approach**. *See also* **intermediate technology**.

aquifer   *See* **artesian basin**.

arable farming   The production of cereal and root crops – as opposed to the keeping of livestock.

arctic circle   *See* **latitude**.

arête   A knife-edged ridge separating two **corries** in a glaciated upland. The arête is formed by the progressive enlargement of corries by **weathering** and **erosion**. *See also* **pyramidal peak**.

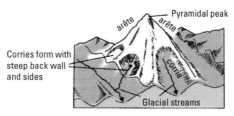

arête

artesian basin   This consists of a shallow **syncline** with a layer of **permeable rock**, e.g. chalk, sandwiched between two impermeable layers, e.g. clay. Where the permeable rock is exposed at the surface, rainwater will enter the rock and the rock will become saturated. This is known as an *aquifer*. Boreholes can be sunk into the structure to tap the water in the aquifer. If there is sufficient pressure of water within the aquifer, the water will rise freely to the surface. The London Basin consists of a shallow syncline formed of a layer of chalk between two layers of clay. London's **water table** dropped considerably during the period of significant industrialization (due to demand by the water companies and industry). However, industry has now relocated elsewhere and the water table is rising, causing problems to building, with deep foundations.

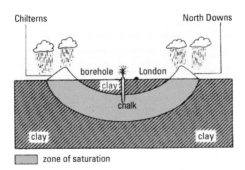

artesian basin

artificial fertilizer   A chemical product containing one or all of the following: nitrogen, potash, phosphates and trace elements, which are derived from petrochemicals, natural phosphate deposits and other industrial sources. In contrast, **organic fertilizers** include animal dung, rotted vegetation (compost) and animal derivatives such as bone meal. Artificial fertilizers, if leached into streams and rivers, can cause problems of **eutrophication**.

artificial fibres   Textile materials, e.g. nylon, rayon, viscose, and polyester, made from hydrocarbons, derived from such materials as oil, **coal** and wood fibres, in contrast with natural fibres such as wool and cotton.

assembly industry   A firm which assembles components into a finished product. For example, the motor-car industry assembles parts such as engines, bodies, wheels, windscreens and electrical components into a final product. The individual components are made by a number of other firms. Assembly industry is distinguished from a manufacturing industry such as steel-making, which uses primary products such as **coal, limestone** and iron ore. *See also* **primary sector**.

asymmetrical fold   Folded **strata** where the two **limbs** are at different angles to the horizontal.

atmosphere   The air which surrounds the Earth, and consists of three layers: the *troposphere* (6 to 10 km from the Earth's surface), the *stratosphere* (50km from

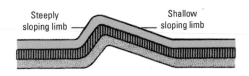

asymmetrical fold

the Earth's surface), and the *mesosphere* and *ionosphere*, an ionised region of rarefied gases (1000km from the Earth's surface). The atmosphere comprises oxygen (21%), nitrogen (78%), carbon dioxide, argon, helium and other gases in minute quantities.

attrition   The process by which a river's **load** is eroded through particles, such as pebbles and boulders, striking each other.

B

backwash   The return movement of seawater off the beach after a wave has broken. *See also* **swash** and **longshore drift**.

balance of trade   *See* **international trade**.

bar graph   A graph on which the values of a certain variable are shown by the length of shaded columns, which are numbered in sequence. *Compare* **histogram**.

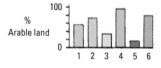

bar graph

barchan   A type of crescent-shaped sand dune formed in desert regions where the wind direction is very constant. Wind blowing round the edges of the dune causes the crescent shape, while the dune may advance in a downwind direction as particles are blown over the crest.

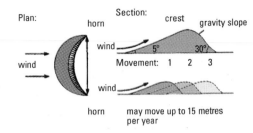

Plan: Section: crest gravity slope
horn
wind 5° 30°
Movement: 1 2 3
wind
wind
horn may move up to 15 metres
per year

barchan

barograph   An aneroid **barometer** connected to an
arm and inked pen which records pressure changes
continuously on a rotating drum. The drum usually
takes a week to make one rotation.

barometer   An instrument for measuring atmospheric
pressure. There are two types, the *mercury barometer*
and the *aneroid barometer*. The mercury barometer
consists of a glass tube containing mercury which
fluctuates in height as pressure varies. The aneroid
barometer is a small metal box from which some of the
air has been removed. The box expands and contracts
as the air pressure changes. A series of levers joined to a
pointer shows pressure on a dial.

barrage   A type of dam built across a wide stretch of
water, e.g. an estuary, for the purposes of water

management. Such a dam may be intended to provide water supply, to harness wave energy or to control flooding, etc. There is a large barrage across Cardiff Bay in South Wales.

barrier beach   A long narrow beach that extends across a bay and which can lead to the formation of a **lagoon** on the landward side.

basalt   A dark, fine-grained extrusive **igneous rock** formed when **magma** emerges onto the Earth's surface and cools rapidly. The Giant's Causeway, Northern Ireland, is composed of basalt. As basalt cools, a hexagonal pattern of jointing may occur as it contracts – hence the hexagonal basalt columns of the Giant's Causeway and of Fingal's Cave, on the island of Staffa, Scotland. A succession of basalt **lava flows** may lead to the formation of a **lava plateau**.

base flow   The water flowing in a stream which is fed only by **groundwater**. During dry periods it is only the base flow which passes through the stream channel.

basin of internal drainage   In certain desert regions there are depressions (sometimes below sea level) from which there is no natural outlet. Intermittent streams drain to the centre of the basin, from which **drainage** occurs by evaporation. Lakes occupying such basins fluctuate in depth and area according to the balance of inflow and evaporation. Extensive salt deposits (e.g.

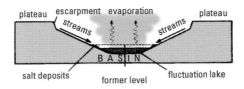

basin of internal drainage

halite, gypsum) may be left behind as lakes evaporate,
e.g. Dead Sea and the Makgadikgadi Pan (Botswana).

batholith   A large body of igneous material intruded
into the Earth's **crust**. As the batholith slowly cools,
large-grained **rocks** such as **granite** are formed.
Batholiths may eventually be exposed at the Earth's
surface by the removal of overlying rocks through
weathering and **erosion**. Dartmoor and Exmoor in
Southwest England are offshoots (bosses) from an
underlying batholith.

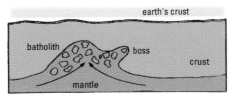

batholith

bay   An indentation in the coastline with a **headland** on either side. Its formation is due to the more rapid **erosion** of softer rocks.

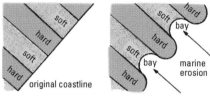

bay

bay bar   A bank of sand or shingle extending as a barrier almost or totally across a **bay**, caused by **longshore drift**.

beach   A strip of land sloping gently towards the sea, usually recognized as the area lying between high and low tide marks.

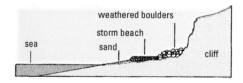

beach

beach deposits Beach deposits may be varied, comprising sand and shingle deposited by the sea, banks of pebbles deposited in storm tides, and boulders which have been weathered out of the **cliffs** behind. Sand is the final product of marine **erosion**, formed largely by **attrition** of the marine **load**.

bearing A compass reading between 0 and 360 degrees, indicating direction of one location from another:

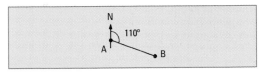

bearing *The bearing from A to B is 110°.*

Beaufort wind scale An international scale of wind velocities, ranging from 0 (calm) to 12 (hurricane).

bedding plane The division between two **strata** of rock, generally indicating the boundary between earlier and later periods of **deposition**.

bergschrund A large **crevasse** located at the rear of a **corrie** icefield in a glaciated region, formed by the weight of the ice in the corrie dragging away from the rear wall as the **glacier** moves downslope (*see figure*).

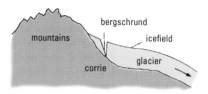

bergschrund

biodiversity   The existence of a wide variety of plant and animal species in their natural environment. Conservationists are concerned that pollution and other human interventions such as intensive agriculture and genetically modified crops are reducing biodiversity and hence the stability of **ecosystems**.

biogas   The production of methane and carbon dioxide, which can be obtained from plant or crop waste. Biogas is an example of a renewable source of energy (*see* **renewable resources, nonrenewable resources**).

biomass   The total number of living organisms, both plant and animal, in a given area.

biosphere   The part of the Earth which contains living organisms. The biosphere contains a variety of **habitats**, from the highest mountains to the deepest oceans.

birth rate The number of live births per 1,000 people in a population per year.

bituminous coal Sometimes called house coal – a medium-quality **coal** with some impurities; the typical domestic coal. It is also the major fuel source for **thermal power stations**.

block faulting The dissection of a region by an extensive system of vertical or semi-vertical **faults**. The faults divide the landscape into a series of blocks which may be relatively uplifted or depressed, to produce a series of **block mountains** (*horsts*) and **rift valleys**.

block faulting

block mountain or horst A section of the Earth's **crust** uplifted by faulting. Mt Ruwenzori in the East African Rift System is an example of a block mountain.

blowhole   A crevice, **joint** or **fault** in coastal rocks,
enlarged by marine **erosion**. A blowhole often leads
from the rear of a cave (formed by wave action at the
foot of a **cliff**) up to the cliff top. As waves break in the
cave they erode the roof at the point of weakness and
eventually a hole is formed. Air and sometimes spray
are forced up the blowhole to erupt at the surface. At
times wave action can be so strong that bursts of
seawater are sent through the blowhole, enlarging it
further, as shown below.

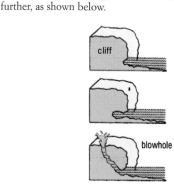

blowhole

blue-collar worker   A worker who is either a manual
worker or who works in potentially dirty conditions.

The term 'blue-collar' derives from the wearing of dark-coloured overalls which show less dirt than light-coloured clothing. *Compare* **white-collar worker**.

blue ice   Heavily compressed ice at the bottom of a **corrie** icefield or **glacier**, the expulsion of air caused by the compression making the ice appear blue. Contrast this with **white ice**.

bluff   *See* **river cliff**.

boulder clay or till   The unsorted mass of debris dragged along by a **glacier** as *ground moraine* and dumped as the glacier melts. Boulder clay may be several metres thick and may comprise any combination of finely ground 'rock flour', sand, pebbles or boulders.

bourne   A spring with a fluctuating origin on the **dip slope** of a chalk escarpment. In wet seasons, the spring may emerge higher up the dip slope, i.e. when the **water table** is higher. Drier weather will cause a drop in the level of the water table and the spring will then emerge further down the dip slope. (*See figure overleaf.*)

The term 'bourne' is commonly applied to streams in the chalk regions of southern England.

BP   *abbrev. for* Before Present, a term used in geological time-scales to denote any time before the present.

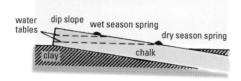

bourne

break-of-bulk point   A point in freight carriage
where goods are offloaded from one mode of transport
and loaded onto another. A port is a major break-of-
bulk point, as is a railhead. Minimizing break-of-bulk
points is one way of reducing costs in transport systems.

breakwater or groyne   A wall built at right angles
to a beach in order to prevent sand loss due to
longshore drift. (*See figure opposite.*)

breccia   Rock fragments cemented together by a
matrix of finer material; the fragments are angular and
unsorted. An example of this is volcanic breccia, which
is made up of coarse angular fragments of **lava** and **crust**
rocks welded by finer material such as ash and **tuff**.

brownfield site   A site for development in an urban
area that has previously been used (for example, the site

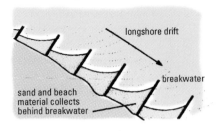

breakwater or groyne

of a demolished factory). *Compare* **greenfield site**.

Burgess model or concentric theory   A model of urban structure formulated by E.W. Burgess in 1923 and based on Chicago, in which five major concentric zones of urban land-use are proposed.

1 – Central business district
2 – Transition zone (factory zone)
3 – Working men's homes
4 – Single family dwellings
5 – Commuter zone

At the core is the **CBD**, with housing zones of varying quality radiating outwards. The further out one goes, the better the housing becomes. This model, together with the **sector model** and **multiple nuclei model**, can provide a basis for looking at city structure in the developed world. *Compare* **shanty town**.

bush fallowing or shifting cultivation   A system of **agriculture** in which there are no permanent fields. For example in the **tropical rainforest**, remote societies cultivate forest clearings for one year and then move on. The system functions successfully when forest **regeneration** occurs over a sufficiently long period to allow the soil to regain its fertility. (*See figure opposite.*)

business park   An out-of-town site accommodating offices, high-technology companies and light industry. Business parks are usually built on a **greenfield site** (one not previously built on), with landscaped grounds and low-density buildings. It is typical for 70% of the site to be open, with grass, flowerbeds and trees, to create a pleasant environment for the workers. Unlike a **science park**, a business park does not have a direct link with a university. Some business parks also contain retail outlets such as hypermarkets and DIY stores, and some even include leisure complexes.

butte   An outlier of a **mesa** in arid regions.

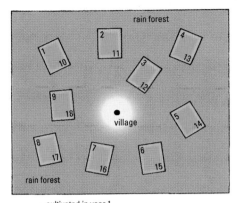

rain forest

village

rain forest

cultivated in year 1

and in year 10

bush fallowing or shifting cultivation

C

caldera   A large crater formed by the collapse of the summit cone of a **volcano** during an eruption. The caldera may contain **subsidiary cones** built up by subsequent eruptions, or a crater lake if the volcano is extinct or dormant.

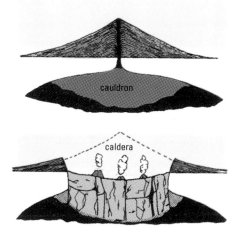

caldera

canyon   A deep and steep-sided river valley occurring where rapid vertical **corrasion** takes place in arid regions. In such an **environment** the rate of **weathering** of the valley sides is slow. If the **rocks** of the region are relatively soft then the canyon profile becomes even more pronounced. The Grand Canyon of the Colorado River in the USA is the classic example.

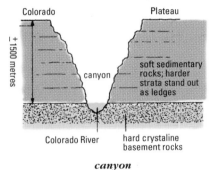

canyon

cap rock   **1.** (also called *fall maker*) A stratum of resistant **rock** at the lip of a **waterfall**.
2. Hard rock protecting the top of a **mesa**.

capital intensive   Relating to an operation in which high productivity is achieved through high investment; for example, **market gardening** is a capital intensive form of **agriculture** identified by high investment in equipment such as greenhouses and heating and irrigation systems. *See also* **labour intensive**.

catchment   **1.** In **physical geography**, an alternative term to **river basin**.
2. In **human geography**, an area around a town or city – hence 'labour catchment' means the area from which an urban workforce is drawn.

cavern   In **limestone** country, a large underground cave formed by the dissolving of limestone by subterranean streams. *See also* **stalactite, stalagmite**.

CBD (Central Business District)   This is the central zone of a town or city, and is characterized by high **accessibility**, high land values and limited space. The visible result of these factors is a concentration of high-rise buildings at the city centre. The CBD is dominated by retail and business **functions**, both of which require maximum accessibility.

census   The collection of information, used in particular to establish the accurate population of a country. In the UK, a questionnaire is sent to all **households** every ten years by the Office of Population Censuses and Surveys.
    Information required by the census includes: the age and sex of the occupants; which people in the household work; where they work and how they travel to work; the number of rooms in the house; the nature

of the **household amenities**; and whether the house has a garage. This information is used to assess the future needs of particular areas and households, e.g. whether new schools or old people's homes are likely to be required in the future. It is illegal not to fill in the census form.

central place A **settlement** offering goods and **services** to a surrounding population.

central place theory Walter Christaller's 1933 model of **settlement** location and distribution.

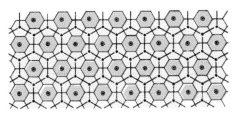

- first order central place

first order sphere of influence

second order central place

second order sphere of influence

other places can be added

central place theory

Christaller envisaged a **settlement hierarchy** in which small **central places** would offer a limited range of everyday goods and services to a small surrounding population, and large central places would offer a large number of goods and services, many of them of a specialized nature, to a large surrounding population. Thus the **sphere of influence** of the large central places would be considerably greater in extent than that of the small central places. Christaller proposed several orders of central places. The figure shows a typical Christaller hierarchy for an idealized and uniform **environment**.

CFCs (Chlorofluorocarbons)   Chemicals used in the manufacture of some aerosols, the cooling systems of refrigerators and fast-food cartons. These chemicals are harmful to the **ozone** layer and their use is being greatly reduced.

chalk   A soft, whitish **sedimentary rock** formed by the accumulation of small fragments of skeletal matter from marine organisms; the rock may be almost pure calcium carbonate. In Britain, chalk occurs in the low hills of the south and east, for example in the Downs and the Chilterns. Due to the **permeable** and soluble nature of the rock, there is little surface **drainage** in chalk landscapes.

chernozem   A deep, rich soil of the plains of southern Russia. The upper **horizons** are rich in lime and other plant nutrients; in the dry **climate** the predominant movement of **soil** moisture is upwards (*contrast* with **leaching**), and lime and other chemical nutrients therefore accumulate in the upper part of the **soil profile**.

cirrus High, wispy or strand-like, thin **cloud** associated with the advance of a **depression**.

clay A soil composed of very small particles of **sediment**, less than 0.002 mm in diameter. Due to the dense packing of these minute particles, clay is almost totally impermeable, i.e. it does not allow water to drain through. Clay soils very rapidly waterlog in wet weather.

cliff A steep rockface between land and sea, the profile of which is determined largely by the nature of the coastal rocks. For example, resistant rocks such as **granite** (e.g. at Land's End) will produce steep and rugged cliffs. The dip of the **strata** influences the gradient of the cliff.

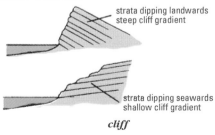

strata dipping landwards
steep cliff gradient

strata dipping seawards
shallow cliff gradient

cliff

climate   The average atmospheric conditions prevailing in a region, as distinct from its **weather.** A statement of climate is concerned with long-term trends. Thus the climate of, for example, the Amazon Basin is described as hot and wet all the year round; that of the Mediterranean Region as having hot dry summers and mild wet winters. *See* **extreme climate, maritime climate**.

climatic change   Fluctuations in the patterns of **climate** over long periods of time. There is indisputable evidence for changes in the world's climate: (a) *fossil records*: fossils of species such as the mammoth have been found in what are now temperate regions; (b) *topographical evidence*: in North Africa, now largely desert, it can be seen where rivers once ran, indicating that the area's rainfall was once much higher; (c) *historical records*: the freezing over of the River Thames was widely reported in the 17th and 18th centuries (*see* **Little Ice Age**); (d) *meteorological records*: records of the weather which exist for many areas of the world and some of which date from early times; comparisons can be made with present-day climate; (e) *geological evidence*: the landscape of, for example, much of Britain has been shaped by the action of glaciers (*see* **Ice Age**).

clint   A block of **limestone**, especially when part of a **limestone pavement,** where the surface is composed of clints and **grykes**.

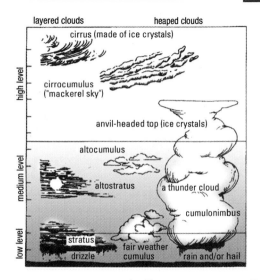

layered clouds   heaped clouds

cirrus (made of ice crystals)

high level

cirrocumulus ("mackerel sky")

anvil-headed top (ice crystals)

medium level

altocumulus

altostratus

a thunder cloud

cumulonimbus

low level

stratus

drizzle

fair weather cumulus

rain and/or hail

cloud

cloud A mass of small water drops or ice crystals formed by the **condensation** of water vapour in the **atmosphere**, usually at a considerable height above the Earth's surface. There are three main types of cloud: **cumulus, stratus** and **cirrus**, each of which has many variations.

coal   A **sedimentary rock** composed of decayed and compressed vegetative matter. Coal in Britain dates from the Carboniferous period (about 350 million years ago), when tropical forest covered large areas of the land.

Coal is usually classified according to a scale of hardness and purity ranging from **anthracite** (the hardest), through **bituminous coal** and **lignite** to **peat**.

coastal marsh   A marsh formed by the growth of a **spit** across a **bay**, and the gradual silting up of the

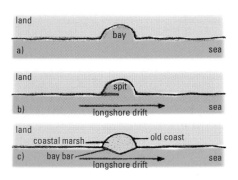

coastal marsh

resulting **lagoon**. A well known example is Romney Marsh in southeast England (*see figure*).

cold front   *See* **depression**.

colonial influence   The consequences of, for example, British colonial activity in many parts of the developing world. There are two broad areas where colonialism has had long-term and continuing influence: in government and administration, and in economic systems. Many former colonies have governments and bureaucracies modelled on institutions of the former colonial power, and this may not always be to the best advantage of the nation concerned. Economically, the majority of former colonies continue to be suppliers of **primary products** to the industrial world, a situation which may well perpetuate underdevelopment.

commercial agriculture   A system of **agriculture** in which food and materials are produced specifically for sale in the market, in contrast to **subsistence agriculture**. Commercial agriculture tends to be **capital intensive**. *See also* **agribusiness**.

Common Agricultural Policy (CAP)   The policy of the European Union to support and subsidize certain crops and methods of animal husbandry. It has been criticized for supporting inefficient farming techniques and also for encouraging overproduction. Minor reforms are under consideration.

common land   Land which is not in the ownership of an individual or institution, but which is historically available to any member of the local community. Common grazing rights still remain in some upland areas of Britain. The traditional English village green, where it survives, is often common land.

communications   The contacts and linkages in an **environment**. For example, roads and railways are communications, as are telephone systems, newspapers, and radio and television. *See also* **route**.

commuter zone   An area on or near to the outskirts of an urban area. Commuters are among the most affluent and mobile members of the urban community and can afford the greatest physical separation of home and work. The commuter zone is thus the outer ring of suburbs (*see* **Burgess model**) and the villages beyond, the latter increasingly peopled by managers, professionals, etc., who travel daily to the city centre. *See* **dormitory settlement**.

comprehensive redevelopment   A policy of clearing substandard housing and completely rebuilding the area to create a new environment. Comprehensive redevelopment usually takes place in **inner city** areas, where conditions of overcrowding may be severe, and where houses are often mingled with industry and business. Comprehensive redevelopment aims to provide houses with good amenities, with areas of open space and parks, away from industry and business. Some schemes have been successful, but many schemes in the 1960s, where high-rise blocks of flats were included, have come in for much criticism.

concealed coalfield A coalfield in which **coal** measures are located at some depth beneath the overlying **strata** (area 2 in the figure). In contrast is the *exposed coalfield* where coal measures outcrop at or near the surface, e.g. the Yorkshire coalfield (area 1 in the figure).

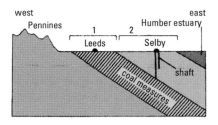

concealed coalfield

concentric theory *See* **Burgess model**.

concordant coastline A coastline that is parallel to mountain ranges immediately inland. A rise in sea level or a sinking of the land cause the valleys to be flooded

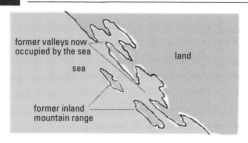

concordant coastline

by the sea and the mountains to become a line of
islands. This has occurred along the coast of Croatia in
the Adriatic sea. (*See figure above.*) *Compare* **discordant
coastline**.

condensation   The process by which cooling vapour
turns into a liquid. **Clouds**, for example, are formed by
the condensation of water vapour in the **atmosphere**.

coniferous forest   A forest of **evergreen** trees such
as pine, spruce and fir. Natural coniferous forests occur
considerably further north than forests of broad-leaved
deciduous species, as coniferous trees are able to
withstand harsher climatic conditions. The **taiga** areas of
the northern hemisphere consist of coniferous forests.
Not all coniferous forests are indigenous to an area as
some have been planted for commercial reasons (*see*

afforestation). It is anomalous that the larch, a species common to many coniferous forests, is a deciduous conifer.

conservation The preservation and management of the natural **environment**. In its strictest form, conservation may mean total protection of endangered species and habitats, as in nature reserves. In some cases, conservation of the man-made environment, e.g. ancient buildings, is undertaken.

continental climate The climate at the centre of large landmasses, typified by a large annual range in temperature, with precipitation most likely in the summer.

continental drift The theory that the Earth's continents move gradually over a layer of semi-molten rock underneath the Earth's **crust**. It is thought that the present-day continents once formed the supercontinent, **Pangaea,** which existed approximately 200 million years ago. *See also* **Gondwanaland, Laurasia** *and* **plate tectonics**.

continental shelf The seabed bordering the continents, which is covered by shallow water – usually of less than 200 metres. Along some coastlines the continental shelf is so narrow it is almost absent. Sunlight easily penetrates the seas of continental shelves, therefore plankton, plants and fish are abundant there.

contour   A line drawn on a map to join all places at
the same height above sea level. Contour heights are
expressed in metres on British Ordnance Survey maps
(e.g. the 1:50,000 series), and the interval between
contours is usually 10 m (the 1:50,000 series) or 5m (the
1:25,000 series). Contours are generally shown
in brown.

contour ploughing   A method of soil **conservation**
whereby ploughing is undertaken along **contours** rather
than with the slope. The effect of this strategy is to
reduce the rate of runoff and thus to retain **soil** that
would otherwise be washed off downslope. The risk of
sheet erosion and gully erosion (*see* **soil erosion**) is
thus reduced.

contract farming   A system in which individual
farmers contract with food-processing firms to produce
a specific crop. The firm will often provide seed and
other inputs at the start of the season, and will purchase
the entire crop at harvest time. Contract farming is
common in East Anglia (or a system where farmers
contract with contract farmers who will farm all the
land for the farmer). *See also* **agribusiness**.

conurbation   A continuous built-up urban area
formed by the merging of several formerly separate
towns or cities. Hence, for example, the West Yorkshire
conurbation includes Leeds, Bradford, Wakefield,

Huddersfield, Halifax and many smaller towns. Several stages can be identified in the growth of conurbations.

Evidence of these processes can be observed, for example, on the Yorkshire coalfield, where small rural **settlements** grew dramatically during the 19th century as steam power gave rise to the expansion of mining and the manufacturing industry. Twentieth-century **urban sprawl** has led to the merging of towns.

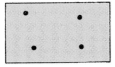

Before 1800 – villages

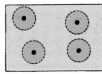

1800 to 1900
Rapid industrialization
and urbanization

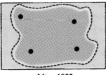

After 1900
Towns merge into a
continuous urban area
– a conurbation

----- limit of built-up areas

conurbation

convection The transfer of energy through a fluid (e.g. air, water, molten material) by movement within the fluid. An increase in the temperature of a fluid causes an increase in its volume and a decrease in its density. This sets up a convection current as hotter fluid rises and displaces the cooler, denser fluid.

Warm air circulates by convection currents in the **atmosphere**. The same process causes movement of liquid rock in the upper layers of the Earth's **mantle** and is responsible for the welling up of **magma** at a constructive plate boundary and the subduction of ocean crust at a destructive plate boundary, resulting in the movement of crustal plates (*see* **plate tectonics**).

cooperative A system whereby individuals pool their **resources** in order to optimize individual gains.

A farming cooperative would, for example, enable farmers to have the use of machinery which could not normally be afforded by individuals. Cooperative systems exist in many developing countries, for example there is the Ujamaa system in Tanzania.

Cooperatives may also facilitate small-scale manufacturing projects in developing countries, such as the production of handicrafts for sale to tourists and for export to the developed world.

core 1. In **physical geography**, the core is the innermost zone of the Earth. It is probably solid at the centre, and composed of iron and nickel.

2. In **human geography**, a **central place** or central region, usually the centre of economic and political activity in a region or nation. John Friedman's *core/periphery model* identifies relationships between growing and stagnant regions during the process of economic development as follows:

(a) *Preindustrial phase:* independent villages with local **spheres of influence**.

(b) *Industrializing phase:* a core region emerges, based on a growing industrial urban centre – the rest of the nation remains an undeveloped **periphery**, supplying labour, food and other **resources** to the core.

(c) *Fully developed phase:* regional urban centres develop to spread the benefits of economic and social progress to the former peripheries.

This is a highly simplified version of a complex model of regional development.

Core *Preindustrial phase.*

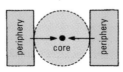

Core *Industrializing phase.*

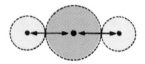

Core Fully developed phase.

corrasion The abrasive action of an agent of **erosion** (rivers, ice, the sea) caused by its **load**, for example the pebbles and boulders carried along by a river wear away the channel bed and the river bank. *Compare* with **hydraulic action**.

corrie, cirque or cwm A bowl-shaped hollow on a mountainside in a glaciated region; the area where a valley **glacier** originates. In glacial times the corrie contained an icefield, which in cross section appears as in figure *a* opposite. The shape of the corrie is determined by the rotational erosive force of ice as the glacier moves downslope (figure *b*).

corrosion **Erosion** by solution action, such as the dissolving of **limestone** by running water.

counter-urbanization The movement of population in economically developed countries from the cities to the rural areas. People move seeking a better quality of

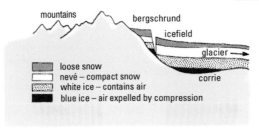

(a) *A corrie in glacial times.*

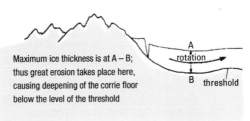

Maximum ice thickness is at A – B; thus great erosion takes place here, causing deepening of the corrie floor below the level of the threshold

(b) *Erosion of a corrie.*

life, even though it frequently means commuting to the cities for work.

crag A rocky outcrop on a valley side formed, for example, when a **truncated spur** exists in a glaciated valley. Crags are exposed to **weathering**, especially by

nivation, and the resulting debris accumulates as **scree** beneath the crag.

crag and tail A feature of lowland **glaciation**, where a resistant rock outcrop withstands **erosion** by a **glacier** and remains as a feature after the **Ice Age**. Rocks of volcanic or metamorphic origin are likely to produce such a feature. As the ice advances over the crag, material will be eroded from the face and sides and will be deposited as a mass of boulder clay and debris on the leeward side, thus producing a 'tail'. An example of a crag and tail is Castle Rock and the Royal Mile, Edinburgh.

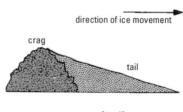

direction of ice movement

crag

tail

crag and tail

crater lake A lake occupying a **caldera** of a dormant or extinct **volcano**.

crevasse A crack or fissure in a **glacier** resulting from the stressing and fracturing of ice at a change in

gradient or valley shape. Crevasses range from small surface cracks to major fractures many metres in depth, and may occur at any angle, although two main orientations are recognized: transverse and longitudinal. Transverse crevasses occur where there is a change of gradient in the valley floor; longitudinal where the valley becomes wider and the ice mass stretches to occupy the broader space.

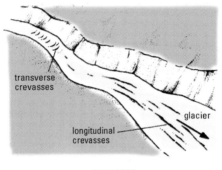

transverse crevasses

glacier

longitudinal crevasses

crevasse

cross section   A drawing of a vertical section of a line of ground, deduced from a map. It depicts the **topography** of a system of **contours**. (*See diagram overleaf.*)

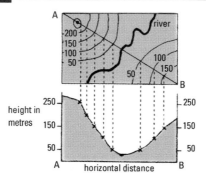

cross section *Map and corresponding cross section.*

crust   The outermost layer of the Earth, representing
only 0.1% of the Earth's total volume. It comprises
continental crust and oceanic crust, which differ from
each other in age as well as in physical and chemical
characteristics. The most ancient parts of the
continental crust are 4000 million years old, whereas the
oldest oceanic crust has an age of less than 200 million
years. The oceanic crust is much thinner than the
continental crust (5–10 km compared with 25–90 km).
The crust, together with the uppermost layer of the
mantle, is also known as the *lithosphere. See also*
plate tectonics.

culvert An artificial drainage channel for transporting water quickly from place to place. Culverting schemes are important in areas prone to flooding. Concrete-lined culverts, with less frictional resistance than a river channel, allow water to flow more freely, thus lessening the risk of flooding during periods of high rainfall.

cumecs Abbreviation for 'cubic metres per second'. *See* **discharge**.

cumulonimbus A heavy, dark **cloud** of great vertical height. It is the typical thunderstorm cloud, producing heavy showers of rain, snow or hail. Such clouds form where intense solar radiation causes vigorous convection.

cumulus A large **cloud** (smaller than a **cumulonimbus**) with a 'cauliflower' head and almost horizontal base. It is indicative of fair or, at worst, showery **weather** in generally sunny conditions.

cut-off *See* **oxbow lake**.

cyclone *See* **hurricane**.

D

dairying   A **pastoral farming** system in which dairy cows produce milk that is used by itself or used to produce dairy products such as cheese, butter, cream and yoghurt.

death rate   The number of deaths per 1,000 people in a population per year.

deciduous woodland   Trees which are generally of broad-leaved rather than **coniferous** habit, and which shed their leaves during the cold season. The larch is a deciduous conifer and is thus the exception to the **evergreen** norm in such trees.

deflation   The removal of loose sand by wind **erosion** in desert regions. It often exposes a bare rock surface beneath.

deforestation   The practice of clearing trees. Much deforestation is a result of development pressures, e.g. trees are cut down to provide land for agriculture and industry. Deforestation in some **Third World** countries had led to severe **soil erosion**, a consequence of which has been **desertification** and eventually famine. It is thought that many of the famines in African countries during recent years have been partly caused by deforestation.

The felling of huge areas of **tropical rainforest** is causing a change in the balance of oxygen to carbon dioxide in the **atmosphere** (*see* **global warming**).

delta A fan-shaped mass consisting of the deposited **load** of a river where it enters the sea. A delta only forms where the river deposits material at a faster rate

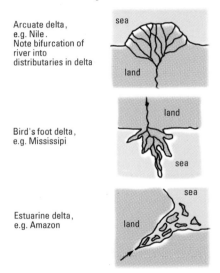

Arcuate delta, e.g. Nile. Note bifurcation of river into distributaries in delta

Bird's foot delta, e.g. Mississipi

Estuarine delta, e.g. Amazon

delta

than can be removed by coastal currents. While deltas may take almost any shape and size, three types are generally recognized, as shown in the figure.

demographic transition theory A model of **population change** which suggests the following pattern of changes over time (*see figure opposite*):

1) *Preindustrial societies:* a high **death rate** due to factors such as low levels of technology and vulnerability to natural disaster; a high **birth rate** to ensure survival of the population.

2) *Industializing societies:* improvements in e.g. nutrition, **communications** and medical facilities, have a quick effect on the death rate, which falls rapidly; the birth rate remains high.

3) *Urban/industrial societies:* in highly developed economies the majority of the population is urban and in paid employment; as a result birth rate falls. Small families are likely partly as a result of a much more mobile population and partly because large families are no longer needed to provide the labour required in pre-industrial, agricultural societies.

The theory can be seen as the history of one nation's population evolution, or a classification applicable to a number of different nations at various stages in the development process. Although the model poses a very generalized and simplified form of reality, it does emphasize the essential link between **population change**, economic development and **urbanization**.

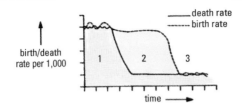

birth/death
rate per 1,000

——— death rate
------- birth rate

time ——→

demographic transition theory

denudation The wearing away of the Earth's surface by the processes of **weathering** and **erosion**.

depopulation A long term decrease in the population of any given area, frequently caused by economic migration to other areas.

deposition The laying down of **sediments** resulting from **denudation**.

depressed region An area of economic stagnation or decline, characterized by high unemployment, out-migration, and declining private and public investment. Depressed regions are often sites of traditional **heavy industry** such as steel-making and engineering; as the **resource** base for such activities contracts, and as demand for their products falls, so the regional economy decays. Parts of northeast England, Merseyside and Clydeside are depressed regions in

Britain. Central government attempts to revitalize
depressed areas through **development area** policy.

depression   An area of low atmospheric pressure
occurring where warm and cold air masses come into
contact, for example, in the case of the northern
hemisphere, along the north polar front (30°–60°N),
where prevailing southwesterly winds (bringing moist,
warm tropical air northwards) meet prevailing
northeasterly winds (bringing cold polar air
southwards). The passage of a depression is marked by
thickening cloud, rain, a period of dull and drizzly
weather and then clearing skies with showers. A
depression develops as in the figures on the right.

desert   An area where all forms of **precipitation** are so
low that very little, if anything, can grow. On average a
desert gets less than 250 mm of rainfall a year, though
some deserts can receive more than 500 mm a year.
However, due to rapid run-off and high rates of
evaporation, this rainwater is unable to support a great
deal of life.

Deserts can be broadly divided into three types,
depending upon average temperatures:

(a) *hot deserts:* occur in tropical latitudes in regions of
high pressure where air is sinking and therefore making
rainfall unlikely. *See* **cloud**. The predominant westerly
winds blow across cold currents before reaching land,
and therefore most rainfall is deposited offshore. The

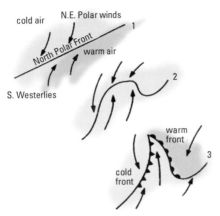

depression *The development of a depression.*

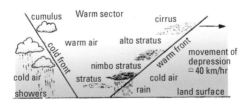

depression *Characteristics.*

daytime temperatures in hot deserts can reach 50°C dropping to –10°C at night, i.e. the diurnal range of temperature is very high.

(b) *temperate deserts:* occur in mid-latitudes in areas of high pressure. They are far inland, so moisture-bearing winds rarely deposit rainfall in these areas. The summer temperatures in these deserts may reach 20°C and will drop to below –20°C during the winter.

(c) *cold deserts:* occur in the northern latitudes, again in areas of high pressure. Very low temperatures throughout the year mean the air is unable to hold much moisture. Summer temperatures may reach 15°C, whilst winter temperatures plunge to below –60°C. *See also* **desertification**.

desertification   The encroachment of **desert** conditions into areas which were once productive. Desertification can be due partly to climatic change, i.e. a move towards a drier climate in some parts of the world (possibly due to **global warming**), though human activity has also played a part through bad farming practices. The problem is particularly acute along the southern margins of the Sahara desert in the Sahel region between Mali and Mauritania in the west, and Ethiopia and Somalia in the east. Low rainfall for several years plus poor management of the land have caused the edge of the Sahara desert to extend southwards.

Although desertification is particularly serious in the Sahel, there are several other parts of the world where it is also a problem. Improved farming practices – irrigation schemes, controlled grazing and **afforestation** – can slow down the process of desertification.

developing countries A collective term for those nations in Africa, Asia and Latin America which are undergoing the complex processes of modernization, **industrialization** and **urbanization**. There is great political and social variation between countries loosely grouped as 'developing', and such terms should be used with caution. *See also* **Third World**.

development area In Britain, a region designated by government as in need of special assistance for economic reconstruction. *See* **depressed region**.

development process The sequence of events by which a nation moves from a predominantly subsistence agricultural economy to one based on **commercial agriculture**, industry and a highly urbanized society.

dew point The temperature at which the **atmosphere**, being cooled, becomes saturated with water vapour. This vapour is then deposited as drops of dew.

differential erosion The unequal **erosion** of interbedded hard and soft rocks, the softer **strata** being worn away more quickly. *See*, for example, **bay**.

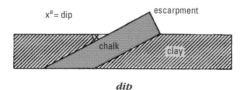

dip

dip   The angle of inclination of **strata** from the horizontal. (*See figure*).

dip slope   The gentler of the two slopes on either side of an escarpment crest; the dip slope inclines in the direction of the dipping **strata**; the steep slope in front of the crest is the **scarp slope**.

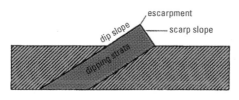

dip slope

discharge   The volume of run-off in the channels of a river basin.

discordant coastline A coastline that is at right angles to the mountains and valleys immediately inland. A rise in sea level or a sinking of the land will cause the valleys to be flooded. A flooded river valley is known as a **ria**, whilst a flooded glaciated valley is known as a **fjord**. *Compare* **concordant coastline**.

discordant coastline

diversification A broadening of an agricultural or industrial product range, in order to reduce dependence on a single, perhaps vulnerable, product. Diversification in **agriculture** is ecologically sound since a more varied **ecosystem** will have a healthier pest-predator complex – uniform wheatfields, for example, are susceptible to explosions of plant-specific pests which require expensive (and perhaps polluting) artificial control. Such **monoculture** also progressively drains the **soil** of nutrients. Diversification is economically sound as an insurance against falling markets for a single product.

doldrums   An equatorial belt of low atmospheric pressure where the **trade winds** converge. Winds are light and variable but the strong upward movement of air caused by this convergence produces frequent thunderstorms and heavy rains.

dormitory settlement   A village located beyond the edge of a city but inhabited by residents who work in that city (*see* **commuter zone**). The populations of dormitory **settlements** are disproportionately large for the number of goods and services available within them (*see* **central place theory**).

drainage   The removal of water from the land surface by processes such as streamflow and infiltration. Drainage can be hastened artificially by the laying of pipes and culverts.

drainage basin   *See* **river basin**.

drift   Material transported and deposited by glacial action on the Earth's surface. *See also* **boulder clay**.

drift mine   A system of mining in which an inclined plane gives access to the ore. In areas of the Yorkshire coalfield, inclined roadways lead to the coalface, allowing free access for plant and machinery.

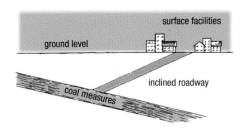

drift mine *Access to coal seam.*

dry valley or coombe A feature of **limestone** and **chalk** country, where valleys have been eroded in what are today dry landscapes. Such valleys may date from a period of more moist **climate**, or from the end of the last **glaciation** when periglacial conditions, specifically a frozen subsoil and bedrock, sealed the otherwise permeable limestone, and thus surface streams existed. (*See* **periglacial features**.)

dyke   **1.** An artificial **drainage** channel.
2. An artificial bank built to protect low-lying land from flooding.
3. A vertical or semi-vertical igneous intrusion occurring where a stream of **magma** (e.g. from a **batholith**) has extended through a line of weakness in the surrounding **rock**. (*See figure overleaf.*)

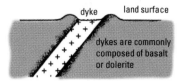

Metamorphosed zone: surrounding
rocks close to intrusion are 'baked'

dyke *Cross section of eroded dyke, showing how
metamorphic margins, harder than dyke or surrounding
rocks, resist erosion.*

E

earthquake   A movement or tremor of the Earth's crust. Earthquakes are associated with plate boundaries (*see* **plate tectonics**) and especially with subduction zones, where one plate plunges beneath another. Here the crust is subjected to tremendous stress. The rocks are forced to bend, and eventually the stress is so great that the rocks 'snap' along a **fault** line. This energy is released as *seismic waves*, originating from the *focus*, i.e. where the earthquake begins.

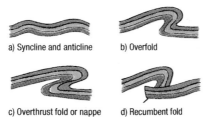

a) Syncline and anticline   b) Overfold

c) Overthrust fold or nappe   d) Recumbent fold

earthquake

eastings The first element of a **grid reference**, as in the following example:

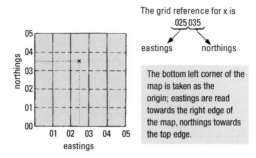

The grid reference for x is

025 035

eastings   northings

The bottom left corner of the map is taken as the origin; eastings are read towards the right edge of the map, northings towards the top edge.

ecology The study of living things, their interrelationships and their relationships with the **environment**.

economies of scale The savings made in industry through mass production, automation and integrated processes. The unit cost of products falls as the quantity manufactured increases; investment in sophisticated manufacturing processes can further reduce unit costs.

ecosystem A natural system comprising living organisms and their **environment**. The concept can be applied at the global scale or in the context of a smaller defined environment, e.g. a woodland, a pond or a marsh. Whatever the scale, the principle of the

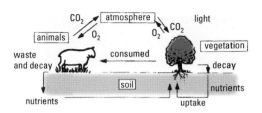

ecosystem

ecosystem is constant: all elements are intricately linked by flows of energy and nutrients, and a change in one element will have effects on the rest of the system. (*See figure above.*)

emigration The movement of population out of a given area or country.

employment structure The distribution of the workforce between the **primary, secondary, tertiary** and **quaternary sectors** of the economy. Primary employment is in **agriculture**, mining, forestry and fishing; secondary in manufacturing; tertiary in the retail, service and administration category; quaternary in information and expertise. One way of looking at the level of development of a nation is to examine its employment structure. Most employment in the **Third World** is primary, while that in the more developed nations is tertiary, with an increasing number of people being employed in the quaternary sector.

enterprise zone An area, usually in the inner city, where special grants and facilities are made available for reconstruction and development.

environment Physical surroundings: **soil**, vegetation, wildlife and the **atmosphere**. Human impact on the environment is a major concern of geographers, especially as human interference can often create problems – for example in causing **pollution**, **soil erosion**, the extinction of species and spread of urban areas. In a broader sense, the term environment is also used to describe the social as well as the physical surroundings of people, such as culture, language, traditions and political systems.

equator *See* **latitude**.

erosion The wearing away of the Earth's surface by running water (rivers and streams), moving ice (**glaciers**), the sea and the wind. These are called the *agents* of erosion.

erratic A boulder of a certain rock type resting on a surface of different geology. For example, at Norber Blocks in Yorkshire, blocks of **granite** rest on a surface of carboniferous **limestone**.

The explanation is glacial: the erratics were picked up

by **glaciers** moving southwards over Britain and deposited at the point where the ice melted. Erratic blocks of this kind may be found at considerable distances from their origin.

escarpment   A ridge of high ground as, for example, the **chalk** escarpments of southern England (the Downs and the Chilterns).

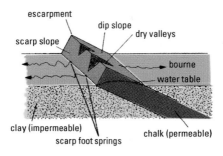

escarpment

esker   A low, winding ridge of pebbles and finer **sediment** on a glaciated lowland. An esker marks the course of a subglacial meltwater stream; as the stream flows, it deposits sediment from the glacier along its course, and this is left behind as a landscape feature at the end of the glacial period.

estuary   The broad mouth of a river where it enters the sea. An estuary forms where opposite conditions to those favourable for **delta** formation exist: deep water offshore, strong marine currents and a smaller **sediment** load (perhaps due to **deposition** upstream, for instance, in a lake).

ethnic group   A group of people with a common identity such as culture, religion or skin colour.

EU (European Union)   Formerly known as the European Community (EC) and the European Economic Community (EEC), this is a group of 15 European countries (Austria, Belgium, Denmark, Finland, France, Germany, Greece, Republic of Ireland, Italy, Luxembourg, Netherlands, Portugal, Spain, Sweden and the United Kingdom) which have agreed to have close political and economic links with each other.

eutrophication   The process whereby an area of water becomes rich in nutrients, causing excessive growth of plants (usually algae) in the springtime. When they decompose, the algae use up the dissolved oxygen in the water, so that wildlife in the pond or lake dies from lack of oxygen.

Eutrophication can be a problem in agricultural areas where fertilizers have leached into the waterways.

evaporation   The process whereby a substance changes from a liquid to a vapour. Heat from the sun evaporates water from seas, lakes, rivers, etc., and this process produces water vapour in the **atmosphere**.

evaporite   A type of **sedimentary rock** formed where salts are precipitated by evaporation in hot climates. Typical evaporites are gypsum and halite (rock salt), which are formed around the margins of lakes in **basins of internal drainage**, e.g. the Dead Sea and the Sea of Galilee.

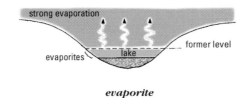

evaporite

evapotranspiration   The return of water vapour to the **atmosphere** by evaporation from land and water surfaces and the **transpiration** of vegetation.

evergreen   A vegetation type in which leaves are continuously present. *Compare* **deciduous woodland**.

exfoliation   A form of **weathering** whereby the outer layers of a **rock** or boulder shear off due to the alternate expansion and contraction produced by diurnal heating and cooling. Such a process is especially active in **desert** regions where, due to clear skies, night temperatures drop considerably below daytime peaks. Isolated boulders may be surrounded by exfoliation debris.

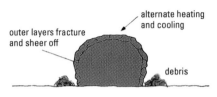

outer layers fracture
and sheer off

alternate heating
and cooling

debris

exfoliation

expanded town   An existing town where growth has been planned in order to accommodate newcomers. Funding will have come from local and national government. Expanded towns are favoured as being less costly than **new towns**. Examples in England include Swindon and Basingstoke.

exponential growth   A rapid form of growth, for example where a population doubles in every unit of time: 2, 4, 8, 16, 32. This can be contrasted with *arithmetic growth*, for example 2, 4, 6, 8, 10 where the growth is the same in every unit of time.

exports   Goods and services sold to a foreign country (*compare* **imports**).

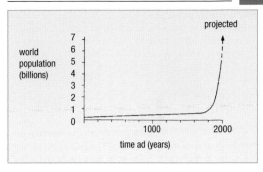

exponential growth

exposed coalfield   *See* **concealed coalfield.**

extensive farming   A system of **agriculture** in which relatively small amounts of capital or labour investment are applied to relatively large areas of land. For example, sheep ranching is an extensive form of farming, and yields per unit area are low. Extensive farming usually occurs at the *margin* of the agricultural system – at a great distance from the market, or on poor land of limited potential. *See also* **Von Thünen theory.**

external processes   Landscape-forming processes such as **weather** and **erosion**, in contrast to internal processes.

extreme climate A climate that is characterized by large ranges of temperature and sometimes of rainfall. In central Asia, for example, hot summers (average 21°C) alternate with very cold winters (average −45°C). Thus the average temperature range is normally more than 60°C. Most rainfall occurs in summer. *Compare* **temperate climate, maritime climate**.

F

false colour image *See* **satellite image**.

fault A fracture in the Earth's crust on either side of which the **rocks** have been relatively displaced. The scale of faulting can vary considerably, from a small surface crack to a fracture of regional extent. Faulting occurs in response to stress in the Earth's crust; the release of this stress in fault movement is experienced as an **earthquake**. *See also* **rift valley**.

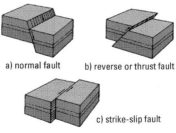

a) normal fault   b) reverse or thrust fault

c) strike-slip fault

fault The main types.

fell Upland rough grazing in a **hill farming** system, for example in the English Lake District. Fell land is sometimes **common land**, i.e. not in the ownership of a single individual or institution. Sheep are grazed on the fells in the summer months and brought down to lower pasture during the winter.

field sketch   A sketch of a landform or landscape made in the field. The value of a field sketch is that it forces the observer to examine carefully the features which are in view. Field sketches are usually simple, yet pick out key geographical features. Careful labelling is an integral part of any field sketch.

fjord   A deep, generally straight inlet of the sea along a glaciated coast. A fjord is a glaciated valley which has been submerged either by a post-glacial rise in sea level or a subsidence of the land.

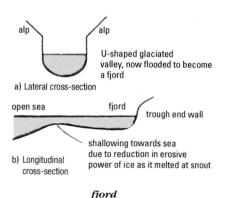

alp   alp

U-shaped glaciated
valley, now flooded to become
a fjord

a) Lateral cross-section

open sea   fjord
                               trough end wall

shallowing towards sea
due to reduction in erosive
power of ice as it melted at snout

b) Longitudinal
cross-section

fjord

flash flood   A sudden increase in river **discharge** and overland flow due to a violent rainstorm in the upper **river basin**.

flood plain   The broad, flat valley floor of the lower course of a river, levelled by annual flooding and by the lateral and downstream movement of **meanders**.

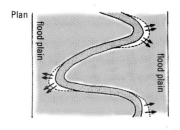

flood plain

flow line   A diagram showing volumes of movement, e.g. of people, goods or information between places. The width of the flow line is proportional to the amount of movement, for example in portraying commuter flows into an urban centre from surrounding towns and villages. (*See figure overleaf.*)

fodder crop   A crop grown for animal feed, either for direct feeding, e.g. turnips, or for making **silage**, as with grass grown for hay.

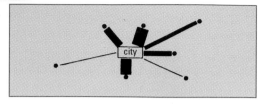

Flow line *Commuter flows into a city.*

fold A bending or buckling of once horizontal rock
strata. Many folds are the result of rocks being
crumpled at plate boundaries (*see* **plate tectonics**),
though **earthquakes** can also cause rocks to fold, as can
igneous **intrusions**. The flat plain of southern England
has been folded into a series of **anticlines** (arch-shaped)
and **synclines** (trough-shaped).

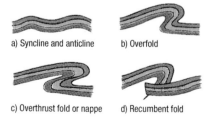

a) Syncline and anticline   b) Overfold

c) Overthrust fold or nappe   d) Recumbent fold

fold

fold mountains   Mountains that have been formed by large-scale and complex folding. Studies of typical fold mountains (the Himalayas, Andes, Alps and Rockies) indicate that folding has taken place deep inside the Earth's **crust** and upper **mantle** as well as in the upper layers of the crust.

Such folding is the result of compression of the continental crust when plates converge (*see* **plate tectonics**). When continental crust and oceanic crust converge, the oceanic crust is carried downwards beneath the continental plate (*subduction*), and the leading edge of the continental crust is compressed and folded upwards to form a mountain range parallel with the coast, such as the Andes. When two continental plates collide, folding and uplifting occurs at both leading edges as the plates fuse together. The Alps and the Himalayas were formed in this way.

food chain   The feeding succession, starting with *producers*, i.e. green plants, and leading through to various *consumers*, i.e. animals. (*See figure overleaf.*)

In practice single food chains do not exist. Several chains interwoven form a food web. *See* **ecosystem**.

footloose industry   Any manufacturing industry with no specific requirements or preconditions for its location. Such industries are neither resource- nor market-orientated, and as such can enjoy a freedom to locate in a wide variety of areas. Footloose industries

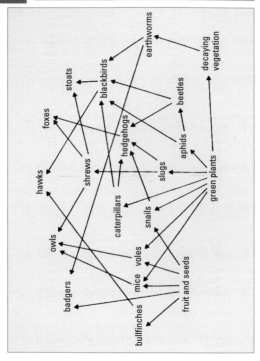

food chain

tend to be found in modern consumer societies in which motorways provide good transport mobility and where there is an industrial trend towards assembling products from components shipped in from a wide variety of sources. Most **light industries** are footloose as they do not have to be located close to **raw materials** or power supplies. The location of footloose industries contrasts with that of traditional **heavy industries** which were restricted to specific areas. Iron and steel plants, for example, were located close to supplies of coal and iron ore.

fossil fuel   Any naturally occurring carbon or hydrocarbon fuel, notably coal, oil, peat and natural gas. These fuels have been formed by decomposed prehistoric organisms.

   The burning of fossil fuels, particularly since the **Industrial Revolution**, has led to problems of **acid rain** in many countries of the world and to increased levels of carbon dioxide (CO_2) in the atmosphere, a factor in the **greenhouse effect**. Most governments are committed to stabilizing or reducing CO_2 emission from burning fossil fuels. Fossil fuels are non-renewable energy resources (*see* **renewable resources, nonrenewable resources**).

free trade   The movement of goods and services between countries without any restrictions (such as quotas, tariffs or taxation) being imposed.

front A boundary between two air masses. *See also* **depression**.

functions The term used for goods and services available in a **central place**; in general, the number and variety of functions offered increase with the size of the **settlement**. Everyday or convenience (low order) functions are available in small settlements, while these plus specialized or durable (high order) functions, are available in large settlements. Small settlements may offer only two or three functions; large settlements many hundreds. *See also* **central place theory**.

G

GDP (gross domestic product)   The total value of the goods and services produced annually by a nation.

gentrification   The process whereby old, run-down houses, often in **inner city** areas, are improved and restored by outsiders who have higher incomes than the local population. Gentrification can change the social structure of a community; as house prices rise, so the original local population are no longer able to afford to rent or purchase the restored properties.

geosyncline   A basin (a large **syncline**) in which thick marine sediments have accumulated.

geothermal energy   A method of producing power from heat contained in the lower layers of the Earth's **crust**. New Zealand and Iceland both use superheated water or steam from geysers and volcanic **springs** to heat buildings and for hothouse cultivation and also to drive steam turbines to generate electricity. In Britain, experiments in Southampton have successfully used geothermally heated water from rocks below the city to heat shops and offices in the city centre. In the long term, scientists are hoping to be able to tap heat from the granite rocks in southwest England. Geothermal energy is an example of a renewable resource of energy (*see* **renewable resources, nonrenewable resources**). It is relatively pollution-free, but hot mineral waters that are not used have to be disposed of carefully to avoid polluting surface drainage systems.

glacial advance   The extension of **ice sheets** and **glaciers** to lower altitudes to cover large areas. This is caused by a cooling of the **climate**.

glacial retreat   A reduction in the area covered by **ice sheets** and **glaciers**, caused by a warming of the **climate**. *Compare* **glacial advance**.

glaciation   A period of cold **climate** during which time **ice sheets** and **glaciers** are the dominant forces of **denudation**.

   The last glaciation ended about 10,000 years ago, and much of Britain's landscape (north of a line drawn approximately between the Thames and the Severn) shows evidence of the effects of ice.

glacier   A body of ice occupying a valley and originating in a **corrie** or icefield. A glacier moves at a rate of several metres per day, the precise speed depending upon climatic and **topographic** conditions in the area in question. The Mer de Glace and Aletsch glaciers in the Alps are present-day examples.

global warming or greenhouse effect   The warming of the Earth's atmosphere caused by an excess of carbon dioxide, which acts like a blanket, preventing the natural escape of heat. This situation has been developing over the last 150 years because of (a) the burning of **fossil fuels**, which releases vast amounts of carbon dioxide into the **atmosphere**, and (b) **deforestation**, which results in fewer trees being available to take up carbon dioxide (*see* **photosynthesis**).

A global rise in temperature is likely to have many consequences. For example, parts of the polar ice sheet would melt, raising sea level, which would result in the flooding of low-lying coastal areas throughout the world. This would cause loss of life, and in addition the loss of farmland and food supplies, housing and industry. It would damage transport networks and power supplies, and would pollute inland water supplies with salt water. In other parts of the world, rising temperatures would cause the development of new areas of desert, making food production difficult in formerly fertile areas. The process of global warming could be slowed down by reduction of fossil fuels, replanting of trees and stricter controls on the destruction of the world's forests.

globalization   The process that enables financial markets and companies to operate internationally (as a result of deregulation and improved communications). **Transnational corporations** now locate their manufacturing in places that best serve their global market at the lowest cost.

GNP (gross national product)   The total value of the goods and services produced annually by a nation, plus net property income from abroad. GNP per capita is the GNP divided by the total population of the country. It is a crude measure as it does not take account of the purchasing power of money, the income distribution in the country, or the real income in developing countries, where many farmers grow food for subsistence rather than cash.

Gondwanaland   The southern-hemisphere supercontinent, consisting of the present South America, Africa, India, Australasia and Antarctica– that split from **Pangaea** *c.*200 million years ago. Gondwanaland is part of the theory of **continental drift**. *See also* **plate tectonics**.

graded profile   The long profile of a river's course that would exist after all irregularities have been removed by **erosion**.

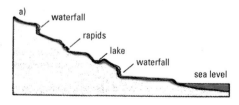

graded profile (a) *Ungraded profile.*

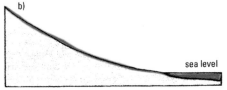

graded profile (b) Graded profile.

gradient **1.** The measure of steepness of a line or slope. In mapwork, the average gradient between two points can be calculated as:

$$\frac{\textit{difference in altitude}}{\textit{distance apart}}$$

Gradient between A and B:

$$\frac{300m - 50m}{800m}$$

$$= \frac{250m}{800m} = \frac{1}{3.2} \qquad \textit{Expressed as 1:3.2}$$

Interpretation: 1 m of ascent for every 3.2 m of horizontal equivalent.

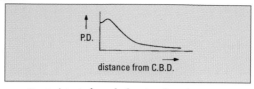

gradient *A typical graph showing the urban population density gradient.*

2. The measure of change in a property such as density. In **human geography** gradients are found in, for example, **population density**, land values and **settlement** ranking.

granite   An **igneous rock** having large crystals due to slow cooling at depth in the Earth's **crust**.

Granite is a common constituent of **batholiths**, and is mainly composed of **quartz**, mica and feldspar minerals.

green belt   An area of land, usually around the outskirts of a town or city on which building and other developments are restricted by legislation. The purpose of such planning law is to attempt to preserve open space and relatively rural **environments** which would otherwise be lost with the advance of **urban sprawl**.

green revolution   The introduction of high-yielding crops of rice and wheat into developing countries. These crops require fertilizers and can only be used by farmers who have appropriate funds. They have helped to increase food production.

greenfield site   A development site for industry, retailing or housing that has previously been used only for agriculture or recreation. Such sites are frequently in the **green belt**.

greenhouse effect   *See* **global warming**.

grid reference   A method for specifying position on a map. *See* **eastings**.

groundwater   Water held in the bedrock of a region, having percolated through the **soil** from the surface. Such water is an important **resource** in areas where **surface runoff** is limited or absent.

gryke   An enlarged joint between blocks of **limestone** (**clints**), especially in a **limestone pavement**. Grykes are progressively enlarged by the solution effect of rainwater.

H

habitat   A preferred location for particular species of
plants and animals to live and reproduce. A habitat may
be small, such as a rock-cranny, or large, such as a
tropical rain forest.

hanging valley   A tributary valley entering a main
valley at a much higher level because of deepening of
the main valley, especially by glacial erosion. During
glaciation, vertical erosion is much greater in the main
valley than in smaller tributary valleys. Small tributary
valleys will only have small glaciers in them, or no
glaciers at all. Once the ice has retreated, the floor of
the main valley lies far below the tributary valleys which
are therefore called 'hanging valleys'.

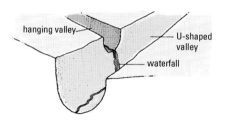

hanging valley

Harris and Ullman model   *See* **multiple nuclei model**.

HDI (human development index)   A measure created by the United Nations to assess the economic and social development of a country. It combines information on life expectancy, education, and the purchasing power of incomes. The index goes from 0 to 1, the highest score representing the most developed.

headland   A promontory of resistant **rock** along the coastline. *See* **bay**.

heave   A horizontal displacement of **strata** at a fault.

heavy industry   Traditional industries using bulky **resources**, e.g. coal mining, iron and steel-making, chemicals and engineering.

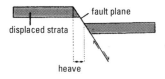

heave

high-technology approach An approach to the **development process** which stresses the role of capital and sophisticated technology. It is argued that investment in large-scale **resource** management schemes (e.g. dams for **hydroelectric power**), and in the industrial sector in general, is the surest way to hasten national development. In Brazil, for example, development priorities have been identified in this way.

Very often, the capital for high-technology schemes in developing countries is provided by developed countries. *Contrast* this with the **intermediate technology** (and **appropriate technology**) approach.

hill farming A system of **agriculture** where sheep (and to a lesser extent cattle) are grazed on upland rough pasture.

In Britain, hill farming occurs on **marginal land** in

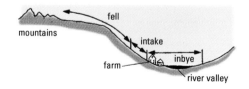

hill farming

upland areas such as the Lake District, Snowdonia and the Scottish Highlands. The typical hill farm comprises three zones: the *inbye*, *intake* and **fell**. The inbye is valley-bottom land, immediately surrounding the farm buildings; it is walled or fenced and may be cultivated for **fodder crops** and sown pasture. The intake extends up the lower slopes of the surrounding **fells** and is an area of sheltered pasture for winter grazing and for lambing. The fell is an extensive area of upland rough grazing to which several farmers may have right of access.

hinterland An area inland from a port, and defined by the limit of the port's **sphere of influence**, for example:

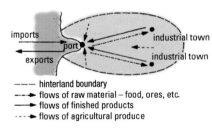

----- hinterland boundary
---→ flows of raw material – food, ores, etc.
——→ flows of finished products
---→ flows of agricultural produce

hinterland

histogram   A graph for showing values of classed data as the areas of bars. *See diagram below. Compare* **bar graph**.

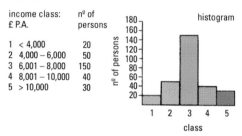

income class: £ P.A.	nº of persons
1 < 4,000	20
2 4,000 – 6,000	50
3 6,001 – 8,000	150
4 8,001 – 10,000	40
5 > 10,000	30

histogram

home region   The area around a person's home. This might be on a small scale, e.g. the area within a short travelling distance, or on a broader scale, several counties. For example, the home region for a person living in London could be regarded as southeast England.

honeypot   A key site in a tourism area that particularly attracts visitors so drawing them into an area they might otherwise not visit.

horizon   The distinct layers found in the **soil profile**. Usually three horizons are identified – A, B and C, as in the figure opposite.

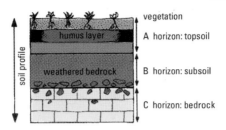

horizon *A typical soil profile.*

The A horizon or **topsoil** contains humus and other vegetable debris. The B horizon or subsoil contains a larger proportion of inorganic material, and receives minerals washed down from the topsoil by the process of **leaching**. The division between the B and C horizons is marked by a zone of decaying bedrock. In reality there will rarely be sharp divisions between zones; the A and B horizons, for example, may be mixed by the activity of worms, burrowing animals or root growth. *See also* **inorganic fraction, organic fraction**.

horticulture The growing of plants and flowers for commercial sale. It is now an international trade, for example, orchids are grown in Southeast Asia for sale in Europe.

household   The number of people living in a single dwelling. This could be one person in a bedsit, or a family of six in a house. A block of flats would be made up of several households. Information about households is required by the ten-yearly national **census**.

household amenities   Utilities in a dwelling such as gas, electricity and running hot and cold water, which are important for everyday life. Older dwellings may have poorer amenities, such as a shared bathroom and an outside toilet, compared with many modern dwellings which have very good amenities, e.g. central heating. *See also* **neighbourhood amenities**.

human geography   The study of people and their activities in terms of patterns and processes of population, **settlement**, economic activity and **communications**. There is no precise definition of such a broad subject, but the basic task of the human geographer is to try to explain distributions of people and their activities. *Compare* **physical geography**.

hunter/gatherer economy   A pre-agricultural phase of development in which people survive by hunting and gathering the animal and plant **resources** of the natural **environment**. No cultivation or herding is involved. Few hunter/gatherer societies survive today, but the few examples we do have include the Wiama Indians and other tribes living in the Amazon rainforest.

hurricane, cyclone or typhoon   A wind of force 12 on the **Beaufort wind scale**, i.e. one having a velocity of more than 118 km per hour. Hurricanes can cause great damage by wind as well as from the storm waves and floods that accompany them.

Hurricanes occur in the belt of the **trade winds** where these winds begin to diminish towards the **doldrums**. Such tropical storms usually arise at the end of summer when the ocean is warmest. This means that the warm, moist air above the ocean is likely to rise, and **condensation** will occur. This releases latent energy (i.e. the heat originally used to evaporate the water) as the air spirals upwards, and which warms the centre, or *eye*, of the storm. This warm, dry air descends, expands and sucks up more moisture, which again is carried upwards. Thus, the hurricane is 'fed' by the warm ocean, and wind speeds increase as the winds circle the eye of the storm. The hurricane will continue as long as it is fed with moisture from the sea and with heat from condensation. This explains why most hurricanes lose force and 'die' when they reach land, though severe damage can occur in coastal areas.

hydraulic action   The erosive force of water alone, as distinct from **corrasion**. A river or the sea will erode partially by the sheer force of moving water and this is termed 'hydraulic action'. For example, a swift river current will undercut the outer bank of a **meander**. The force of waves breaking against a **headland** will compress the air in rock crevices and thereby cause the surrounding **rock** to shatter.

hydroelectric power   The generation of electricity by turbines driven by flowing water. Hydroelectricity is most efficiently generated in rugged **topography** where a head of water can most easily be created, or on a large river where a dam can create similar conditions. Whatever the location, the principle remains the same – that water descending via conduits from an upper storage area passes through turbines and thus creates electricity.

hydrograph   *See* **storm hydrograph**.

hydrological cycle   The cycling of water through sea, land and **atmosphere**.
 The amount of fresh water available in the system is small: 97% of all the water on the Earth's surface is salt, and of the remaining 3% which is fresh, most is locked in **ice sheets**. (*See figure on facing page.*)

hydrosphere   All the water on Earth, including that present in the **atmosphere** as well as in oceans, seas, **ice sheets**, etc.

hygrometer   An instrument for measuring the relative humidity of the **atmosphere**. It comprises two thermometers, one of which is kept moist by a wick inserted in a water reservoir. Evaporation from the wick reduces the temperature of the 'wet bulb' thermometer, and the difference between the dry and the wet bulb temperatures is used to calculate relative humidity from standard tables.

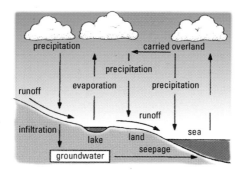

hydrological cycle

I

Ice Age A period of **glaciation** in which a cooling of **climate** leads to the development of **ice sheets, ice caps** and valley **glaciers**. The most recent Ice Age was the Quaternary glaciation, ending about 10,000 years ago. *Compare* **Little Ice Age**.

ice cap A covering of permanent ice over a relatively small land mass, e.g. Iceland.

ice fall An area of fractured ice in a **glacier** where a change of **gradient** occurs.

ice sheet A covering of permanent ice over a substantial continental area such as Antarctica.

igneous rock A **rock** which originated as **magma** (molten rock) at depth in or below the Earth's **crust**. Igneous rocks are generally classified according to crystal size, colour and mineral composition; intrusive and extrusive types are also recognized.
1) **Batholith:** a large body of magma intruded into the Earth's crust; this cools slowly at depth to form igneous rocks with large crystals such as **granite**.
2) **Dyke:** vertical or semi-vertical sheet of igneous rock; a minor intrusion compared with a batholith. Dolerite is a common dyke rock.

3) **Sill:** horizontal or semi-horizontal minor intrusion; sills and dykes exploit lines of weakness (e.g. **joints, bedding planes, faults**) in the crustal rocks.

4) **Lava flow:** extrusive igneous rocks are those which reach the Earth's surface via some form of volcanic eruption. Such rocks have small crystals due to rapid cooling on the Earth's surface. **Basalt** is a common example.

The terms volcanic, *hypabyssal* and **plutonic** are also used to describe igneous rocks: **volcanic rocks** are those which are extruded onto the surface; hypabyssal rocks are those of intrusions such as sills and dykes; and **plutonic rocks** are those of deep intrusions such as batholiths.

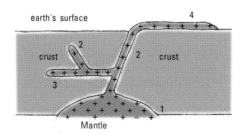

igneous rock

immigration   The movement of people into a country or region from other countries or regions.

impermeable rock   A rock that is non-porous and therefore incapable of taking in water or of allowing it to pass through between the grains. *Compare* **impervious rock**. *See also* **permeable rock**.

impervious rock   A non-porous rock with no cracks or fissures through which water might pass. An **impermeable rock** such as **granite** may be pervious due to the presence of **joints**.

imports   Goods or services bought into one country from another (*compare* **exports**).

inbye   *See* **hill farming**.

industrial estate   A purpose-built facility for the location of new industry, often situated at the edge of an urban area where land is more readily available and where there is less congestion. Motorway intersections or access points are favoured locations for industrial estates owing to the important role of road transport today. *See also* **business park** *and* **science park**.

industrial inertia   The tendency for an industry to retain original locations even though such locations may no longer be optimum. Steel-making, for example, continues at Sheffield, even though the ores upon which the industry was originally based are long since worked out. The woollen **textile industry** continues to be concentrated in West Yorkshire, even though local factors such as swift streams for water power and, later, **coal** for steam power are no longer relevant. Industries retain their original locations in this way because the costs of relocating are likely to be high. In addition, large amounts of capital are tied up in factory facilities; a skilled labour force may have developed in the region; economies of **agglomeration** and scale may have been established locally. For such reasons it may be cheaper to maintain the original location and to import raw materials from elsewhere. In the case of Sheffield, for example, scrap metal is now the major raw material for the steel industry.

industrial location   The optimum location for an industry is one where the costs of transport are minimized, with respect to both **raw materials** and finished products. The standard theory of industrial location is that of Alfred Weber's *locational triangle:* an industry may locate anywhere within the triangle, depending on the balance of transport costs relating to the three factors.

Various examples can be envisaged, e.g. an industry with bulky raw materials which are expensive to transport will locate close to factor 1 (*fig. a)*).

According to the theory, an industry such as printing and publishing will locate close to the market, since distribution costs are the most important consideration (*fig. b*)).

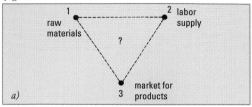

a)

locational triangle

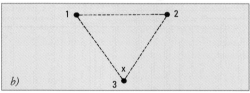

b)

An industry for which all transport costs are similar may locate anywhere in the triangle. In reality, of course, the locational decision is considerably more complex: factors such as the availability of power and the suitability of the land will also be taken into account, as will less tangible issues such as managers' personal preferences and outlook. Government intervention, for example through **development area** policy, may also affect the locational decision. *See* **iron and steel industry** for an example of shifting location over time.

Industrial Revolution   The period in Britain's history from approximately 1780 to 1900 when the invention and application of industrial processes led to the establishment of the world's first urban/industrial society. The invention of the steam engine provided the key to a number of crucial industrial innovations in mining, transport and manufacture. Prior to the Industrial Revolution, successes within agriculture had been due to increased mechanization and efficiency, plus greater profits. This released labour to work in the new industrial cities. Throughout the 19th century large-scale **migration** to urban areas continued, and it was during this period that the major industrial concentrations were established, for example in West Yorkshire, South Lancashire, the West Midlands and Central Scotland.

industrialization   The development of industry on an extensive scale, as happened during the **Industrial Revolution** in Western Europe. Some people identify industrialization with economic development, but for some nations other methods of development may be more appropriate. The term 'industrializing' is sometimes used to describe nations midway through the development process, and **preindustrial** and **post-industrial** nations are identified by the same measure. *See also* **newly industrialized countries**.

infiltration   The gradual movement of water into the ground. The rate at which infiltration occurs is dependent upon two main factors: whether the soil is already wet and saturated, and the porosity of the soil, i.e. whether the water can move easily between the pores of the soil. Water infiltrates quickly through a sandy soil, but slowly through a clay soil.

informal economy   Employment frequently found in **Third World** cities, characterized by irregular hours and wages, limited equipment and machinery, and often operating outside the law. Examples of informal jobs would include taxi drivers who transport materials around the city for business people, shoe shiners, and traders selling goods from trays, e.g. sweets, chewing gum, fruit. In some Third World cities it is estimated that the informal sector employs between 40% and 60% of the working population.

infrastructure   The basic structure of an organization or system. The infrastructure of a city includes, for example, its roads and railways, schools, factories, power and water supplies and drainage systems.

inner city   The ring of buildings around the **CBD** of a town or city. These buildings are usually terraced houses of the early industrial period (many of which may have been modernized – see **gentrification**), blocks of high and low-rise flats (to replace terraced houses which have become dilapidated), as well as offices and small **light industries**. The buildings in the inner city tend to be high-density, i.e. many buildings in a relatively small area.

inorganic fraction   That proportion of the **soil** which is composed of **rock** and mineral fragments and particles deriving from the **weathering** of bedrock. *Compare* **organic fraction**. *See also* **horizon**.

intake   *See* **hill farming**.

integrated steelworks   An industrial site where all iron and steel manufacturing processes are carried out 'under one roof': iron manufacture, steel-making, steel rolling and processing, and alloy manufacture.

intensive farming   A system of **agriculture** where relatively large amounts of capital and/or labour are invested on relatively small areas of land. An example of intensive farming is **market gardening**, where large investment, in the form of greenhouses, fertilization, **irrigation** and heating systems, occurs on small holdings of land close to urban centres. In such a case yields per unit area are high. The small land-holdings reflect the cost of land close to market. *See* **Von Thünen theory** and **agribusiness**. *Compare* **extensive farming**.

interception   *See* **rainfall interception**.

interglacial   A warm period between two periods of **glaciation** and cold **climate**. The present interglacial began about 10,000 years ago.

interlocking spurs   Obstacles of hard **rock** round
which a river twists and turns in a V-shaped valley.
Erosion is pronounced on the concave banks, and this
ultimately causes the development of spurs which
alternate on either side of the river and interlock.

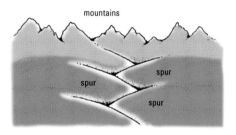

interlocking spurs A V-shaped valley with
interlocking spurs.

intermediate technology   Equipment and facilities
of simple, cheap, practical design directly relevant to the
immediate needs of a community in a **developing country**.
For example, rivers may be regulated by a series of small
earthen dams to provide local flood control and water
for **irrigation**. Such dams can be built with local labour
and equipment, can easily be repaired and are cheap. In

contrast, the **high-technology approach** of sophisticated dams for **hydroelectric power** and industrial development could involve foreign capital, equipment and expertise, and might not be appropriate for a developing country's needs. Intermediate technology, being labour intensive rather than capital intensive, does not leave the country with a big burden of debt in conditions of scarce capital. *See also* **appropriate technology**.

internal processes   Landscape-forming processes which originate below the Earth's surface in the movements connected with **plate tectonics**: folding, faulting and **vulcanicity**.

international trade   The exchange of goods and services between countries. Ideally, a country should aim to export more than it imports, thus establishing a favourable *balance of payments*. **Developing countries** tend to import more than they export, thus incurring large debts to developed countries.

intrusion   A body of **igneous rock** injected into the Earth's **crust** from the **mantle** below. *See* **dyke, sill, batholith**.

inward investment   Financial investment in a country by a **transnational corporation**. The investment may be used to develop industrial or agricultural projects in developing countries and may provide much-

needed employment opportunities. The investment may
also be made in developed countries, with the host
government providing incentives (e.g. tax benefits or a
funding contribution) to the transnational corporation.
Such investments are usually made in regions requiring
new economic activity. However, the outputs of such
projects are usually intended for the transnational
corporation's country of origin.

ionosphere   *See* **atmosphere**.

iron and steel industry   The extraction of iron from
ore and the manufacture of steel, which together are a
key element in the heavy industrial structure of a nation.
In Britain, the location of the iron and steel industry is
an example of changing optimum location over time.
The total pattern of location has three elements. Firstly,
there are the coalfield locations such as Sheffield and
Motherwell, dating from the early establishment of iron
and steel manufacture using **coal** and blackband iron
ore, found in the coal measures, as **raw materials**.
Secondly, there are the orefield locations such as
Scunthorpe (and formerly Corby), developed in the
mid-20th century to exploit the iron ore deposits of
Jurassic rocks in eastern England. By this stage the steel-
making process depended less on coal, owing to the
development of the electric arc furnace and, in blast
furnaces, better designs which required less coal to

produce iron. Thirdly, there are the most recently established coastal locations of the iron and steel industry, for example at Port Talbot in South Wales and on Teesside in northeast England. These locations reflect the dominant role today of imported ores from places such as Scandinavia and Canada. Thus the three locational elements of the overall distribution of the iron and steel industry reflect the three stages of its evolution. Steel-making in general is in decline, and several inland steelworks in the UK have closed (e.g. Consett in Co. Durham, Ravenscraig in Lanarkshire and Corby in Northamptonshire).

irrigation   A system of artificial watering of the land in order to grow crops. Irrigation is particularly important in areas of low or unreliable rainfall. Some of the oldest methods of irrigation were developed along the river Nile and consisted of simple devices such as the shadoof and the Archimedean Screw. Some of these methods are still used in the **Third World** today. The modern method of irrigation (often associated with countries of the developed world where more capital is available for investment in large-scale projects) is to build a dam across a river and create a reservoir. Water is then removed from the reservoir at times of need by a system of pipes and channels. In parts of Britain where rainfall may be unreliable, farmers use hose pipes and sprinklers, e.g. in the **arable farming** regions of East Anglia.

isobar   A line joining points of equal atmospheric pressure, as on the meteorological map below.

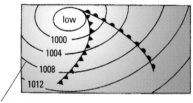

low
1000
1004
1008
1012

isobar, indicating atmospheric pressure in millibars

isobar

isohyet   A line on a meteorological map joining places of equal rainfall.

isotherm   A line on a meteorological map joining places of equal temperature.

J

joint A vertical or semi-vertical fissure in a
sedimentary rock, contrasted with roughly horizontal
bedding planes. In **igneous rocks** jointing may occur as
a result of contraction on cooling from the molten state.
Joints should be distinguished from **faults** in that they
are on a much smaller scale and there is no relative
displacement of the rocks on either side of the joint.
Joints, being lines of weakness, are exploited by
weathering.

K

kame A short ridge of sand and gravel deposited from the water of a melted glacier.

karst topography An area of **limestone** scenery where **drainage** is predominantly subterranean. It is named after a region in Slovenia.

kettle hole A small depression or hollow in a glacial **outwash** plain, formed when a block of ice embedded in the outwash deposits eventually melts, causing the **sediment** above to subside.

L

labour-intensive   Denoting a system of **agriculture** or industry where labour (rather than capital) forms the major input. For example, rice cultivation in much of Southeast Asia is a labour-intensive activity with very little mechanization. As nations develop, capital tends to replace labour in certain sectors of the economy and labour intensity decreases.

laccolith   An igneous **intrusion**, domed and often of considerable dimensions, caused where a body of viscous **magma** has been intruded into the **strata** of the Earth's **crust**. These strata are buckled upwards over the laccolith.

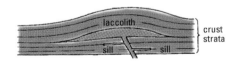

laccolith

lagoon   **1.** An area of sheltered coastal water behind a **bay bar** or **tombolo**.
2. The calm water behind a coral reef. (*See figure overleaf.*)

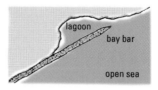

lagoon

lahar   A landslide of volcanic debris mixed with water down the sides of a volcano, caused either by heavy rain or the heat of the volcano melting snow and ice.

land breeze   A breeze occurring at night when the sea is warmer than the land and when, as a result, pressure is lower over the sea. During the day opposite conditions prevail: the land is relatively warmer and the pressure gradient is from land to sea; a sea breeze then occurs. Such conditions arise because the land heats up and cools down more quickly than the sea.

land reform   The process of redistributing land, especially in developing nations where current circumstances may be against fair access to land and against agricultural improvement. For example, in parts of Latin America, the traditional **land tenure** system is made up of large commercial estates (known as *estancias*) and small peasant holdings. Often, the estate occupies the best farming land whilst the peasants farm in difficult conditions such as on valley sides.

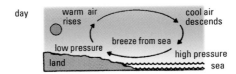

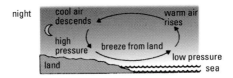

land breeze *Formation of a sea breeze.*

This inequality has led to pressure for land reform, sometimes by violent revolution. The aim of recent land-reform schemes has been to provide typical farming families with security, reasonable farming land, and an incentive to improve productivity. Concern for land reform occurs not only in situations of unequal distribution as described above, but also in traditional rural economies where land is allocated through the chieftainship, often in small and fragmented parcels, as occurs in parts of Africa. Here, the necessary reforms would be consolidation of holdings, the provision of secure tenure, and some facility for the progressive farmer to improve productivity. However, clumsy land reform may do more harm than good: land in many rural societies is much more than a commodity – it is an integral part of the culture and heritage.

land tenure   A system of land ownership or allocation.

land use   The function of an area of land. For example, the land use in rural areas could be farming or forestry, whereas urban land use could be housing or industry. Sometimes land uses can come into conflict; for example, the quarrying of limestone in a **national park** would be in conflict with the aims of **conservation** and public access within the park.

landform   Any natural feature of the earth's surface, such as mountains or valleys.

land-value gradient   The decline of average land values per unit area with increasing distance from the **CBD**. Whilst this is usually true, some suburban areas with good **accessibility** may prove to be good locations for business and retail centres. The land-value gradient may show peaks within suburban areas. (*See figure.*)

lapse rate   The rate at which temperature changes with altitude. In the troposphere, it is usual for temperature to decrease with altitude at an average rate of 0.6°C per 100 metres. In certain conditions of very

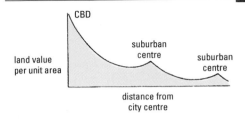

land-value gradient

still air a temperature inversion happens, with temperature increasing with altitude. Such an inversion can trap pollutants close to the surface of the Earth. Beyond the *tropopause*, in the upper part of the **stratosphere**, temperature increases with height. *See also* **atmosphere**.

laterite A hard (literally 'brick-like') soil in tropical regions caused by the baking of the upper **horizons** by exposure to the sun. Laterite occurs often as a result of mismanagement of the tropical **environment**, for example by clear-felling of **tropical rain forest**, or by a decrease in the **regeneration** cycle in shifting agriculture.

Laterite has low agricultural potential; it is difficult to cultivate with simple tools and has a low nutrient content owing to the removal of forest and the consequent reduction in nutrient cycling.

latitude Distance north or south of the equator, as measured by degrees of the angle at the Earth's centre:

e.g.

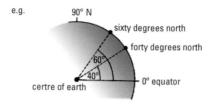

latitude

Laurasia The northern hemisphere supercontinent, consisting of the present North America, Europe and Asia (excluding India), that split from **Pangaea** *c.*200 million years ago. Laurasia is part of the theory of **continental drift**. *See also* **plate tectonics**.

lava **Magma** extruded onto the Earth's surface via some form of volcanic eruption. Lava varies in viscosity (*see* **viscous lava**), colour and chemical composition. Acidic lavas tend to be viscous and flow slowly; basic lavas tend to be nonviscous and flow quickly. Commonly, **lava flows** comprise basaltic material, as for example in the process of sea-floor spreading (*see* **plate tectonics**).

lava flow A stream of **lava** issuing from some form of volcanic eruption. *See also* **viscous lava**.

lava plateau   A relatively flat upland composed of layer upon layer of approximately horizontally bedded lavas. An example of this is the Deccan Plateau of India.

leaching   The process by which soluble substances such as mineral salts are washed out of the upper soil layer into the lower layer by rain water.

least-cost location   The optimum location for an industry. The least-cost location may change over time with transport developments, changes in the sources of raw materials and changing markets. The least-cost location for iron and steel manufacture, for example, has shifted from inland coalfields to coastal sites as imported **raw materials** have increased in importance. *See also* **industrial location**.

levée   The bank of a river, raised above the general level of the **flood plain** by **sediment** deposition during flooding. When the river bursts its banks, relatively coarse sediment is deposited first, and recurrent

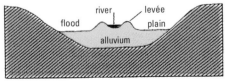

levée

flooding builds up the river's banks accordingly.

Note that continuous deposition on a river bed raises the entire channel above the general level of the flood plain.

light industry   Industry on a relatively small scale, as contrasted with **heavy industry**. Light industry includes industries such as the manufacture of electrical goods, printing and publishing, distribution trades and clothing manufacture.

Light industry is generally to be found on purpose-built estates, often on the edge of an urban area where congestion is less and **communications** are good. Light industry is generally cleaner and less polluting than heavy industry, and is organized mainly for the manufacture of consumer durables.

lignite   A soft form of **coal**, harder than **peat** but softer than **bituminous coal.**

limb   One element of a **fold**, with respect to either an **anticline** or a **syncline.**

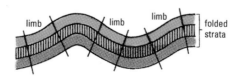

limb *Limbs in a section of folded strata.*

limestone   Calcium-rich **sedimentary rock**
formed by the accumulation of the skeletal matter of
marine organisms. The most favourable conditions
for limestone formation comprise a warm, clear,
shallow sea.

Many limestones contain fossils: shells and skeletal
traces are common. Being both soluble and **permeable**,
limestone gives rise to a dry surface landscape with
characteristic features of underground **drainage**, as
found in the Malham district of Yorkshire, where the
carboniferous limestone, approximately 350 million
years old, is the major element of the landscape.

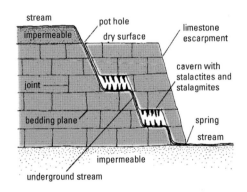

limestone

limestone pavement   An exposed **limestone** surface on which the **joints** have been enlarged by the action of rainwater dissolving the limestone to form weak carbonic acid. These enlarged joints, or **grykes**, separate roughly rectangular blocks of limestone called **clints**.

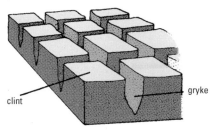

gryke

clint

limestone pavement

link   The connection between two **nodes** in a communication **network**.

Little Ice Age   A period from 1550 to 1850 characterized by extended cold seasons, heavy **precipitation** and the expansion of **glaciers**. It is thought that this may have occurred because of an absence of sunspot activity at this time. The reasons for this are yet to be explained.

load   The **sediment** transported by the agents of **erosion** – rivers, moving ice and the sea. The size and volume of load transported depends upon the power of the transporting medium. In a river system, for example, more load is carried in times of high **discharge**. A river's load comprises material rolled or bounced along the bed, material carried in suspension, and material carried in solution. The finest sediment is carried the greatest distances and may contribute to the formation of, for example, a **delta**.

location   The position of population, settlement and economic activity in an area or areas. Location is a basic theme in **human geography**.

location triangle   *See* **industrial location**.

loess   A very fine **silt** deposit, often of considerable thickness, transported by the wind prior to **deposition**. In northern China, for example, there are extensive loess deposits which have been carried by wind from the arid plateau lands of Central Asia. Loess is extremely porous, and the surface is consequently dry. When irrigated, loess can be very fertile and, consequently, high **yields** can be obtained from crops grown on loess deposits.

longitude   A measure of distance on the Earth's surface east or west of the Greenwich Meridian, an imaginary line running from pole to pole through Greenwich in London. Longitude, like **latitude**, is measured in degrees of an angle taken from the centre of the Earth. Lines of longitude are examples of *Great Circles*; a Great Circle is any circle drawn on the Earth's surface, the centre of which coincides with the centre of the Earth. The Great Circle route between two points on the Earth's surface is the shortest possible route. The only line of latitude which is a Great Circle is the **equator**.

The precise location of a place can be given by a **grid reference** comprising longitude and latitude. *See also* **map projection**.

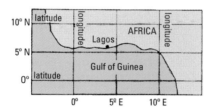

Longitude *A grid showing the location of Lagos, Nigeria.*

longshore drift   The net movement of material along a beach due to the oblique approach of waves to the shore. Beach deposits move in a zig-zag fashion, as shown in the figure. Longshore drift is especially active on long, straight coastlines. The movement of beach

rapidly silt up and it becomes separated from the river. In time, the entire oxbow lake will silt up, pioneer vegetation will invade, and the lake will eventually disappear.

ozone   A form of oxygen found in a layer in the **stratosphere**, where it protects the Earth's surface from ultraviolet rays. There has been recent concern about holes appearing in the ozone layer over the polar regions. There are many possible reasons why these holes might occur, but it is known that the ozone layer is damaged by the chlorine released by **CFCs**. An increase in the ultraviolet rays reaching the Earth could result in an increase in skin cancers, and damage to crops, vegetation, livestock and wildlife in countries near the polar regions.

P

Pangaea The supercontinent or universal land
mass in which all continents were joined together
approximately 200 million years ago. The theory of
Pangaea's existence was devised by Alfred Wegener in
1912. *See* **continental drift**.

pastoral farming A system of farming in which the
raising of livestock is the dominant element. In the
commercial context this may be, for example, dairy
farming in Britain or sheep rearing in Australia, while
subsistence pastoralism occurs in many parts of the
Third World. *See also* **nomadic pastoralism**.

peasant agriculture The growing of crops or
raising of animals, partly for subsistence needs and
partly for market sale. Peasant agriculture is thus an
intermediate stage between subsistence and commercial
farming.

peat Partially decayed and compressed vegetative
matter accumulating in areas of high rainfall and/or
poor **drainage**.
  In Britain, peat occurs in the upland areas of the north
and west, forming *blanket-bog* over much of the
Pennines, and in low-lying parts of East Anglia. Peat **soil**
results from the limited breakdown of fallen vegetation in

anaerobic conditions: total breakdown into humus cannot occur in waterlogged, airless conditions. Upland peat is acidic and infertile due to **leaching**; lowland peat has a higher nutrient status and is more useful for **agriculture**.

The removal of lowland peat for use as a garden fertilizer is currently a controversial issue. The large-scale removal of peat disrupts the natural ecosystem, and **conservation** measures are now being looked at to preserve many of Britain's peat lands.

pedestal rock   *See* **Zeugen**.

peneplain   A region that has been eroded until it is almost level. The more resistant rocks will stand above the general level of the land.

per capita income   The **GNP** (gross national product or national income) of a country divided by the size of its population. It gives the average income per head of the population if the national income were shared out equally. Per capita income comparisons are used as one indicator of levels of economic development.

percolation   The movement of water through soil pores and rock crevices.

periglacial features   A periglacial landscape is one which has not been glaciated *per se*, but which has been affected by the severe **climate** prevailing around the ice margin. Intensive **nivation** is characteristic of periglacial **environments**, as is *solifluction*, a process whereby thawed surface soil creeps downslope over a permanently frozen **subsoil** (*permafrost*). Much of the Canadian **tundra** and Siberian heartland is affected by such periglaciation.

periphery   A remote and/or underprivileged region as in the core/periphery model (*see* **core**). Such regions are generally lacking in resources and offer little development opportunity, and as such are the last to be integrated into the national development process.

permafrost   The permanently frozen subsoil that is a feature of areas of **tundra**.

permeable rock   Rock through which water can pass via a network of pores between the grains. Examples are sandstone and chalk. *Compare* **pervious rock**. *See also* **impermeable rock**.

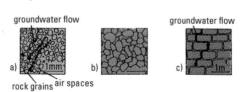

groundwater flow

rock grains air spaces

groundwater flow

permeable rock   *(a) Permeable rock, (b) impermeable rock, (c) pervious rock.*

pervious rock   Rock which, even if non-porous, can allow water to pass through via interconnected **joints, bedding planes** and **fissures**. An example is **limestone**. *Compare* **permeable rock**. *See also* **impervious rock**.

pH   A measure of acidity/alkalinity. A pH value of 7.0 is regarded as neutral, while pH values of less than 7.0 indicate acidic conditions. pH values greater than 7.0 indicate increasingly alkaline conditions. pH tests are used to assess the acidity of soil and of water. *See* **acid rain**. The optimum soil pH for cereal growth is about 6.5.

photosynthesis   The process by which green plants make carbohydrates from carbon dioxide and water, and give off oxygen. Photosynthesis balances **respiration**. However, the burning of **fossil fuels** has greatly increased the amount of carbon dioxide in the atmosphere, and **deforestation** has seriously reduced the number of trees available to release oxygen, so that the natural balance has been lost. *See* **global warming**.

physical geography   The study of our **environment**, comprising such elements as geomorphology, hydrology, pedology, meteorology, climatology and biogeography.

pie chart   A circular graph for displaying values as proportions (*see overleaf*):

The journey to work: mode of transport.
(Sample of urban population)

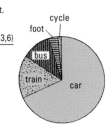

Mode	No	%	Sector° (% x 3,6)
Foot	25	3,2	11,5
Cycle	10	1,3	4,7
Bus	86	11,2	40,4
Train	123	15,9	57,2
Car	530	68,5	246,3
Total	774	100	360
		percent	degrees

pie chart

plantation agriculture   A system of **agriculture**
located in a tropical or semi-tropical **environment**,
producing commodities for export to Europe, North
America and other industrialized regions. Coffee, tea,
bananas, rubber and sisal are examples of plantation
crops.

   Plantation agriculture is distinctive in that it is a
form of **commercial agriculture** located in a generally
subsistence or peasant environment: it is an extension
of the commercial agriculture of the developed world
into a mainly **Third World** environment. Some
plantations are run and financed by **transnational
corporations** and the profits from such operations are

generally channelled back to Europe or North America. As a result, many plantations are seen as institutions of **neocolonialism**. There is a worry that plantations often take up valuable farmland, growing commodities required by the richer countries. Thus more valuable local crops are forced on to poorer land. In areas where unemployment is high, plantations have traditionally paid low wages. On a more positive note, many plantation operators provide such facilities as housing, education and health care for their workers, as well as a plot of land. But it is difficult to avoid the conclusion that plantation agriculture in its traditional form is unacceptable in view of contemporary development priorities in Third World nations.

plate tectonics The theory that the Earth's **crust** is divided into seven large, rigid plates, and several smaller ones, that are moving relative to each other over the upper layers of the Earth's **mantle**. *See* **continental drift**. **Earthquakes** and volcanic activity occur at the boundaries between the plates.

There are three types of plate boundary:
(a) *constructive plate boundary* – where new oceanic crust is formed on either side of the boundary as **magma** wells up (*see* **convection**), increasing the width of the ocean bed by a few centimetres per year (*sea-floor spreading*) and thus helping to push the continents on either side further apart.

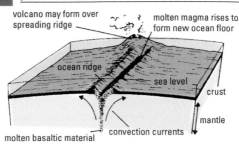

a) Constructive

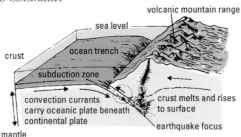

b) Destructive

plate tectonics

(b) *destructive plate boundary* – where the oceanic crust is gradually carried down into the mantle (*subduction*), producing great stresses that cause major earthquakes.

Volcanic eruptions occur at the surface as molten rock rises up through the overlying plate. A destructive plate boundary is marked by an *oceanic trench*; the trenches in the Pacific, about 10,000 m deep, are the deepest places on Earth. Where the oceanic plate is subducted beneath a continental plate, the edge of the continental plate becomes compressed and folded. *See* **fold mountains**.

(c) *conservative plate boundary* – where crustal material is being neither created nor destroyed. Instead the plates slide past each other.

plucking A process of glacial **erosion** whereby, during the passage of a valley **glacier** or other ice body, ice forming in cracks and fissures drags out material from a **rock** face. This is particularly the case with the backwall of a **corrie**.

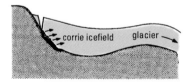

plucking

plug The solidified material which seals the vent of a **volcano** after an eruption. A volcanic plug is thus responsible for the build-up of pressure which may result in an explosive eruption at some later stage. **Viscous lavas** produce the most effective plugs and, because of their resistance to **erosion**, volcanic plugs tend to stand out in the landscape when softer surrounding material has been worn away. A well known example is Edinburgh Castle Rock.

plunge pool   *See* **waterfall**.

plutonic rock   **Igneous rock** formed at depth in the Earth's **crust**; its crystals are large due to the slow rate of cooling. **Granite**, such as is found in **batholiths** and other deep-seated intrusions, is a common example.

podzol   The characteristic **soil** of the **taiga** coniferous forests of Canada and northern Russia. Podzols are leached, greyish soils: iron and lime especially are leached out of the upper horizons, to be deposited as *hardpan* in the B **horizon**.

pollution   Environmental damage caused by improper management of **resources**, or by careless human activity.

In the countries of the 'first world' (*see* **Third World**) progress has been made in the control of some of the worst causes of chemical pollution, such as persistent agricultural chemicals and heavy metals. Smoke control legislation has led to cleaner air in cities (*see* **smog**). However, emissions of sulphur dioxide from the burning of **fossil fuels** by industry, and of nitrous oxides from vehicle exhausts, remain high as governments and industries are reluctant to spend the large amounts of money necessary to reduce them. These pollutants are then blown elsewhere by the **prevailing winds**. Emissions from Britain, for example, are blown over Scandinavia, where the resulting **acid rain** has led to the poisoning of entire **ecosystems**: many lakes are now devoid of fish; forest growth is stunted; and public water supplies require calcification.

In the formerly communist countries of eastern Europe and the USSR, no controls were exerted over chemical pollution by industry or by vehicle exhaust emissions because production was the priority of the governments concerned. Their financial resources were invested in outdated industrial plant that produce enormous quantities of highly polluting emissions. Damage to plants, fish and animals was very serious in some areas, and the incidence of respiratory diseases in the human population of the industrial towns was distressingly high. It will take many years and huge amounts of money to reduce pollution levels, and meanwhile the damage to people and to the environment will continue.

In many **developing countries**, also, production is seen as the priority, and the reduction of pollution as a luxury that the countries cannot afford.

Besides chemical and radioactive pollution (caused by nuclear accidents, as happened at the Chernobyl nuclear power station in the Ukraine in 1987), which may be life-threatening, there are 'nuisance' pollutions such as *noise pollution* by low-flying aircraft and the *visual pollution* of quarries, rubbish dumps, etc.

population change   The increase of a population, the components of which are summarized in the figure on the right.

population density   The number of people per unit area. Population densities are usually expressed per square kilometre, and can range from less than one in remote, inhospitable regions, to many hundreds in urban areas or on highly productive agricultural land.

population distribution   The pattern of population location at a given **scale**. At the global scale, population distribution shows a concentration in specific areas, for example in parts of Asia and Europe, and is sparse in others, such as in the polar regions and the desert areas.

population explosion   On a global **scale**, the dramatic increase in population during the 20th century. The graph overleaf shows world **population growth**.

   The first 1000 million was reached by about 1820; the second by 1930; the third by 1960; and the fourth by about 1975. The world population was estimated to be 6,168 million in 2000 and 9,800 million by 2050 (according to United Nations figures). The greatest growth is expected in Africa. In 1950 8.9% of the

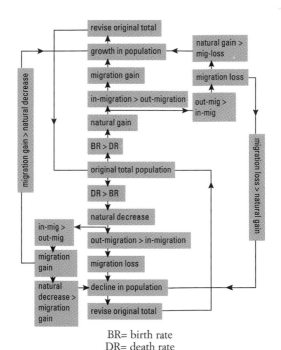

BR= birth rate
DR= death rate

population change

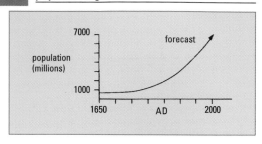

population explosion

world's population was in Africa. It is estimated that, by 2050, 21.8% of the population will live there.

The major cause behind the increasing growth rate is a world-wide fall in **death rate**, resulting from a complex of development factors such as improved nutrition and health care, and better **communications**.

See **demographic transition**.

population growth   An increase in the population of a given region. This may be the result of natural increase (more births than deaths) or of in-migration, or both.

population migration   *See* **migration**.

population pyramid   A type of **bar graph** used to show population structure, i.e. the age and sex

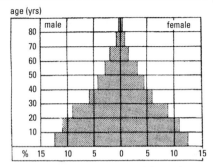

a) population pyramid *Pyramid for India, showing high birth rates and death rates.*

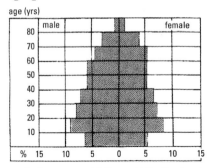

b) population pyramid *Pyramid for England and Wales, showing low birth and death rates.*

composition of the population for a given region or nation.

A population pyramid has data for males plotted on one side and data for females on the other.

The shape of a population pyramid is a useful indicator of the stage of development reached by a nation (*see* **demographic transition theory**) and can also indicate government policy on population planning and the impact of major events such as wars. (*See diagrams on page 171.*)

post-industrial   Characteristic of an economy that is no longer based on **heavy industry** but is increasingly dominated by microelectronics, automation, and the service and information sectors. See **tertiary sector, quaternary sector**.

pothole   **1.** A deep hole in limestone, caused by the enlargement of a **joint** through the dissolving effect of rainwater. The Yorkshire Pennines contain classic examples of potholes, such as Gaping Gill near Ingleton, Yorkshire. Surface streams may disappear down certain potholes, which are then known as *sinkholes* or *swallow holes*, to become underground watercourses.
2. A hollow scoured in a river bed by the swirling of pebbles and small boulders in eddies.

precipitation   Water deposited on the Earth's surface in the form of e.g. rain, snow, sleet, hail and dew. *See* **orographic rainfall**.

preindustrial   A term used to describe the early
stages of the **development process**. Largely agricultural
economies in which the foundations for development
are being established, such as agricultural extension
projects and improved **communications**, would be
described as preindustrial. The implication of such
terminology is that **industrialization** can be equated
with development; while this has been true in recent
history, it may be that alternative models will emerge as
resource shortage and **pollution** become recognized
globally.

prevailing wind   The dominant wind direction
of a region. In northwest Europe, for example,
southwesterly winds predominate, i.e. occur more
frequently than those from any other direction.

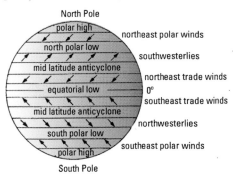

Prevailing winds are named by the direction from which they blow.

primary sector   That sector of the national economy which deals with the production of primary materials: **agriculture**, mining, forestry and fishing. Primary products such as these have had no processing or manufacturing involvement. The total economy comprises the primary sector, the **secondary sector**, the **tertiary sector** and the **quaternary sector**. The distribution of employment between these four sectors is a measure of the state of development of the nation: the early stages of the development process are marked by a concentration of the labour force in the primary sector, especially in agriculture, whilst the progress of development is characterized by increasing employment in, firstly, the secondary, and, later, the tertiary and quaternary sectors. It is also possible to observe regional variations within a country, e.g. the southeast of Britain contains a greater proportion of tertiary and quaternary workers than certain northern areas of Britain. A mining area contains a large number of primary workers, and an industrial region contains a large number of secondary workers.

Many **developing countries** are characterized by the employment of a large percentage of their working population in the primary sector, especially in **agriculture**. This is due partly to a lack of investment in manufacturing – many primary products are exported for processing to the developed nations. For example, a cash crop such as cocoa from Ghana is exported in its raw state to developed countries where it is manufactured into chocolate. The profits made by a developed nation on the sale of the finished chocolate are much greater than those made by the developing country on the raw cocoa.

primary source   *See* **secondary source**.

pull factors   *See* **migration**.

pumped storage   Water pumped back up to the storage lake of a **hydroelectric power** station, using surplus 'off-peak' electricity. The water can then be used again for power generation.

push factors   *See* **migration**.

pyramidal peak   A pointed mountain summit resulting from the headward extension of **corries** and **arêtes.** Under glacial conditions a given summit may develop corries on all sides, especially those facing north and east. As these erode into the summit, a formerly rounded profile may be changed into a pointed, steep-sided peak. The Matterhorn in the Alps is a classic example of this. (*See figure overleaf.*)

pyroclasts   Rocky debris emitted during a volcanic eruption, usually following a previous emission of gases

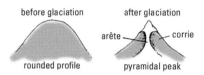

before glaciation

after glaciation

arête

corrie

rounded profile

pyramidal peak

pyramidal peak

and prior to the outpouring of **lava** – although many
eruptions do not reach the final lava stage. Pyroclasts
may comprise lumps of solidified lava from previous
eruptions, as found in a volcanic **plug** or chunks of
country rock, i.e. the crustal material close to the
volcanic vent, or finer debris such as ash and dust.
The largest pyroclasts are *volcanic bombs*, pieces of
debris that may weigh up to several tonnes.
Pebble-sized debris is referred to as *lapilli*.

Q

quality of life   The level of wellbeing of a community and of the area in which the community lives.

quartz   One of the commonest minerals found in the Earth's **crust**, and a form of silica (silicon+oxide). Most **sandstones** are composed predominantly of quartz.

quartzite   A very hard and resistant **rock** formed by the metamorphism of **sandstone**.

quaternary sector   That sector of the economy providing information and expertise. This includes the microchip and microelectronics industries. Highly developed economies are seeing an increasing number of their workforce employed in this sector. *Compare* **primary sector, secondary sector, tertiary sector**.

R

rain gauge An instrument used to measure rainfall. Rain passes through a funnel into the jar below and is then transferred to a measuring cylinder. The reading is in millimetres and indicates the depth of rain which has fallen over an area. Rain gauges are normally read once or twice a day. (*See picture on the right*)

rain shadow *See* **orographic rain**.

rainfall interception The process by which plants and trees break the force of precipitation. Rain lands on the surfaces of the vegetation and fills the hollows. It then runs down the trunks and stems and drips from the leaves on to the ground, where it can sink in gently, instead of running off the surface.

The density and type of vegetation in an area are important in determining the speed at which rainwater moves through the landscape. Different types of vegetation intercept water to a greater or lesser degree. Coniferous forest, for example, intercepts 58% of the rain falling upon it; tall grasses intercept about 27%.

Where large tracts of natural vegetation have been removed from an area, splash erosion and sheet erosion occur (*see* **soil erosion**). Where soil on a slope is not consolidated by a network of roots, it can be loosened by rain and slip off, often with tremendous force, as a

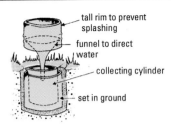

tall rim to prevent splashing

funnel to direct water

collecting cylinder

set in ground

rain gauge

landslide. This has been a big problem in the Himalayas and areas of tropical rain forest, due to **deforestation**.

In urban areas, the lack of vegetation is compensated for by an artificial drainage system.

raised beach   *See* **wave-cut platform**.

range   The maximum distance a consumer is prepared to travel in order to purchase a good or service in a central place (*see* **central place theory**). The range of low-order, everyday goods such as bread, newspapers and daily groceries is very short. Consequently, journeys for these goods are frequent and people will only travel short distances for them. On the other hand, the range of high-order, specialized goods is much greater and people will therefore travel longer distances to purchase these items, and their journeys will be less frequent.

rapid   An area of broken, turbulent water in a river channel, caused by a stratum of resistant **rock** that dips downstream (*see figure*). The softer rock immediately upstream and downstream erodes more quickly, leaving the resistant rock sticking up, obstructing the flow of the water. *Compare* **waterfall**.

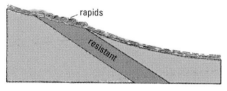

rapids

resistant

rapid

raw materials   The **resources** supplied to industries for subsequent manufacturing processes, for example agricultural products, minerals and timber, are **raw materials**. Many **primary products** are used as raw materials.

regeneration   Renewed growth of, for example, forest after felling. Forest regeneration is crucial to the long-term stability of many **resource** systems, from **bush fallowing** to commercial forestry.

region   An area of land which has marked boundaries or unifying internal characteristics. Geographers may identify regions according to physical, climatic, political, economic or other factors. The Sahara desert is an example of a climatic region, while southeast England (London and the Home Counties) is an economic region.

regional development   A planning policy made by a country to overcome regional differences in income, wealth, education, medical facilities, transport, etc. Such divisions between regions must be minimized if countries are to achieve balanced development. Regional development plans can take many forms; for example, improvements in infrastructure (i.e. the construction of facilities that lay the foundations for further agricultural or industrial development – new roads, power supplies, etc.), resettlement schemes (such as the Indonesian transmigration scheme), or resource development (e.g. mineral exploitation).

Regional inequalities tend to be much more pronounced in **developing countries**, therefore such plans tend to be on a larger and more complex scale than those of developed countries.

rejuvenation   Renewed vertical **corrasion** by rivers in their middle and lower courses, caused by a fall in sea level, or a rise in the level of land relative to the sea.

The point at which downcutting recommences (*knickpoint*) may be marked by a **waterfall**. **Meanders** will become *incised* into the **flood plain**, and **river terraces** may be created.

relative humidity   The relationship between the actual amount of water vapour in the air and the amount of vapour the air could hold at a particular temperature. This is usually expressed as a percentage. Relative humidity gives a measure of dampness in the **atmosphere**, and this can be determined by a **hygrometer**.

relief   The differences in height between any parts of the Earth's surface. Hence a relief map will aim to show differences in the height of land by, for example, **contour** lines or by a colour key.

relief rainfall   *See* **orographic rainfall**.

renewable resources   Resources that can be used repeatedly, given appropriate management and conservation. Water and timber are examples of these. Solar energy and wind energy are also renewable resources. *Compare* **non-renewable resources**.

representative fraction   The fraction of real size to which objects are reduced on a map; for example, on a 1:50,000 map, any object is shown at 1/50,000 of its real size.

reserves   Resources which are available for future use. The world has reserves of **fossil fuels** which, if used with careful management, could last for many years. However, many of our fossil fuel and mineral reserves have been used up at a rapid rate by the developed countries, and more controlled use is required in the future to conserve reserves.

resource   Any aspect of the human and physical **environments** which people find useful in satisfying their needs.

respiration   The release of energy from food in the cells of all living organisms (plants as well as animals). The process normally requires oxygen and releases carbon dioxide. It is balanced by **photosynthesis**.

retail industry   The people and organizations involved in the selling of goods to the public, usually in shops. Goods coming from manufacturers are distributed to the retail industry by *wholesalers*. A *retail park* is a large shopping centre built on the edge of a town to which consumers travel by car.

revolution   The passage of the Earth around the sun; one revolution is completed in 365.25 days. Due to the tilt of the Earth's axis ($23\frac{1}{2}°$ from the vertical), revolution results in the sequence of seasons experienced on the Earth's surface. (*See figure overleaf.*)

ria   A submerged river valley, caused by a rise in sea level or a subsidence of the land relative to the sea. For example, a postglacial rise in sea level may lead to coastal submergence. (*See figure on page 185.*)

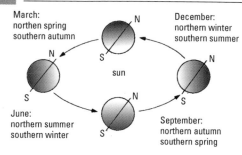

March:
northen spring
southern autumn

December:
northern winter
southern summer

N

sun

June:
northern summer
southern winter

September:
northern autumn
southern spring

revolution *The seasons of the year.*

In the example on the right, a rise in sea level of 25 metres would create a ria. Note the winding plan; contrast with **fjord**.

The seaboard of southwest Ireland is a classic ria coastline. *See also* **discordant coastline**.

ribbon lake   A long, relatively narrow lake, usually occupying the floor of a U-shaped glaciated valley. A ribbon lake may be caused by the *overdeepening* of a section of the valley floor by glacial **abrasion**. Such a situation would occur, for example, where softer or weakened **rocks** outcrop on the valley floor. Alternatively, a ribbon lake may be dammed back by a terminal or recessional **moraine**. Many of the lakes of the English Lake District – for example Windermere, Coniston Water, Wastwater – are ribbon lakes.

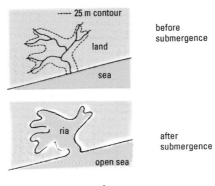

----- 25 m contour

land

sea

before
submergence

ria

open sea

after
submergence

ria

Richter scale   A scale of **earthquake** measurement that
describes the magnitude of an earthquake according to the
amount of energy released, as recorded by **seismographs**.

rift valley   A section of the Earth's **crust** which has
been downfaulted. The **faults** bordering the rift valley
are approximately parallel. There are two main theories
related to the origin of rift valleys. The first states that
tensional forces within the Earth's crust have caused a
block of land to sink between parallel faults. The second
theory states that compression within the Earth's crust
has caused faulting in which two side blocks have risen
up towards each other over a central block.

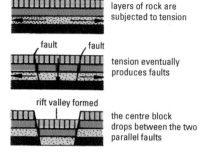

layers of rock are
subjected to tension

tension eventually
produces faults

the centre block
drops between the two
parallel faults

rift valley

The river Rhine flows through a rift valley in Europe, with the Vosges mountains to the west and the Black Forest to the east.

The most complex rift valley system in the world is that ranging from Syria in the Middle East to the river Zambesi in East Africa.

river basin   The area drained by a river and its tributaries, sometimes referred to as a **catchment** area. (*See figure on the right.*)

river cliff or bluff   The outer bank of a **meander**. The cliff is kept steep by undercutting since river **erosion** is concentrated on the outer bank. See meander and **river's course**.

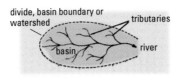

river basin

river's course   The route taken by a river from its
source to the sea. There are three major sections: the
upper course, the middle course and the lower course.
The characteristics of these stages are shown in the
figures below and on the next page.

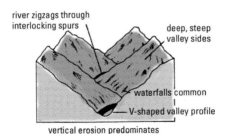

river's course Upper course.

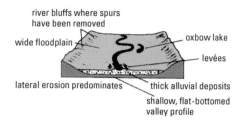

river bluffs where spurs
have been removed

wide floodplain

oxbow lake

levées

lateral erosion predominates

thick alluvial deposits

shallow, flat-bottomed
valley profile

river's course *Lower course.*

river terrace   A platform of land beside a river. This
is produced when a river is **rejuvenated** in its middle or
lower courses. The river cuts down into its **flood plain**,
which then stands above the new general level of the
river as paired terraces.

If a river on a flood plain is rejuvenated several times,
a series of paired terraces can develop.

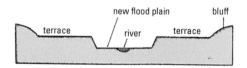

new flood plain

bluff

terrace

river

terrace

river terrace *Paired river terraces above a flood
plain.*

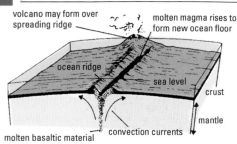

volcano may form over
spreading ridge

molten magma rises to
form new ocean floor

ocean ridge

sea level

crust

mantle

molten basaltic material   convection currents

a) Constructive

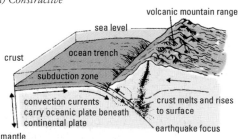

volcanic mountain range

sea level

crust

ocean trench

subduction zone

convection currents
carry oceanic plate beneath
continental plate

crust melts and rises
to surface

mantle

earthquake focus

b) Destructive

plate tectonics

(b) *destructive plate boundary* – where the oceanic crust
is gradually carried down into the mantle (*subduction*),
producing great stresses that cause major earthquakes.

roche moutonnée   An outcrop of resistant **rock** sculpted by the passage of a **glacier**.

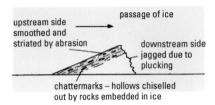

upstream side smoothed and striated by abrasion

passage of ice

downstream side jagged due to plucking

chattermarks – hollows chiselled out by rocks embedded in ice

roche moutonnée

rock   The solid material of the Earth's **crust**. *See* **igneous rock, sedimentary rock, metamorphic rock**.

rotation   The movement of the Earth about its own axis. One rotation is completed in 24 hours. Due to the tilt of the Earth's axis, the length of day and night varies at different points on the Earth's surface. In the northern midsummer, for example, the situation illustrated in the figure overleaf prevails. At the equator, there is a 12-hour day and a 12-hour night. North of 66½°N there is continuous daylight; south of 66½°S there is continuous night. Days become longer with increasing latitude north; shorter with increasing latitude south. The situation is reversed during the northern midwinter (= the southern midsummer).

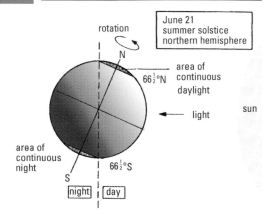

rotation The tilt of the Earth at the northern summer and southern winter solstice.

route   The course taken between starting point and destination. For example, the route of the M1 motorway from London to Leeds is via Leicester and Sheffield. The study of routeways can be important to geographers in establishing the **accessibility** of an area or examining the effects of a new road on the **environment**.

rural depopulation   The loss of population from the countryside as people move away from rural areas towards cities and **conurbations**. Most industrial nations have experienced a shift of population from the countryside to the city as the economy has evolved from an agricultural into a predominantly urban-industrial system. Rural communities tend to lose their most dynamic members as, generally, the migrating population consists chiefly of the younger, most active, most progressive individuals.

rural–urban migration   The movement of people from rural to urban areas. *See* **migration** and **rural depopulation**. *Compare* **counter-urbanization**.

S

sandstone A common **sedimentary rock** deposited by either wind or water.

Sandstones vary in texture from fine -to coarse-grained, but are invariably composed of grains of **quartz**, cemented by such substances as calcium carbonate or silica.

satellite image An image giving information about an area of the Earth or another planet, obtained from a satellite. Instruments on an Earth-orbiting satellite, such as Landsat, continually scan the Earth and sense the brightness of reflected light. When the information is sent back to Earth, computers turn it into *false colour images* in which built-up areas appear in one colour (perhaps blue), vegetation in another (often red), bare ground in a third, and water in a fourth colour, making it easy to see their distribution and to monitor any changes, for example, as a result of **deforestation**. Satellite photography can be used to measure the depth of shallow water, to monitor the health and development of forestry and agricultural crops, and as an aid in oil and mineral exploration.

savanna The grassland regions of Africa which lie between the **tropical rainforest** and the hot **deserts**. Thus in the states of West Africa, for example, a range of vegetation exists from south to north: tropical rainforest to hot desert, via a transition zone of savanna grassland.

Near the forest margin, the savanna region comprises extensive woodland with occasional expanses of grassland; near the desert margin, the vegetation is sparse, thorny scrub. Commercial **pastoral farming** is the dominant land use of the central savannas, while **nomadic pastoralism** prevails towards the desert margin. In South America, the *Llanos* and *Campos* regions are representative of the savanna type.

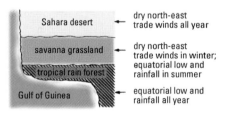

savanna The position of the savanna in West Africa.

scale   The size ratio represented by a map; for example, on a map of scale 1:25,000, the real landscape is portrayed at 1/25,000 of its actual size.

scarp slope   The steeper of the two slopes which comprise an **escarpment** of inclined **strata**. *Compare* **dip slope**.

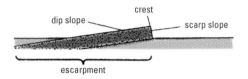

scarp slope

scatter graph   A graph showing the scores of a set of individuals or items on two variables. The line of best fit runs through the points so that the sum of the squares of the vertical distances (or offsets) from the points on to the line is minimal (*see figure a*)). The purpose of the graph is to indicate visually the general trend of the data, i.e. the degree of relationship, or correlation, between the two variables applying to that set of individuals or items. When a unique position for the line cannot be found, because of the purely random distribution of points, as shown in figure b), then no relationship exists and the correlation value is zero.

  The correlation may be positive: the scores on both variables increase or decrease together (figure a)); or negative: the scores on one of the variables increase while those of the other decrease (figure c)).

Displays of points may take many other forms. For example, figure d) shows clustering in which the correlation of each cluster is zero, but all the individuals of each have common characteristics not shared with those of the other clusters.

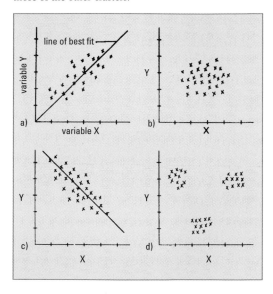

scatter graph

science park   A site accommodating several companies involved in scientific work or research. Science parks are linked to universities and tend to be located on **greenfield** and/or landscaped sites. *Compare* **business park**.

scree or talus   The accumulated **weathering** debris below a **crag** or other exposed rock face. Larger boulders will accumulate at the base of the scree, carried there by greater momentum.

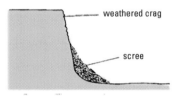

scree or talus

sea breeze   *See* **land breeze**.

sea-floor spreading   *See* **plate tectonics**.

secondary sector   The sector of the economy which comprises manufacturing and processing industries, in contrast with the **primary sector** which produces **raw materials**, the **tertiary sector** which provides **services**, and the **quaternary sector** which provides information.

An example of the secondary sector is the **iron and steel industry**. *See* **heavy industry**.

secondary source   A supply of information or data that has been researched or collected by an individual or group of people and made available for others to use; **census** data is an example of this. A *primary source* of data or information is one collected at first hand by the researcher who needs it; for example, a traffic count in an area, undertaken by a student for his or her own project.

sector model or Hoyt model   A model of urban structure developed by H. Hoyt in 1939 and based on an analysis of the land-use patterns of 142 American cities.

   The model differs from the **Burgess model** in allowing sectorial development along major lines of communication. Thus, zone 2 (industry) and its related workers' housing zone (3) will develop along, for example, a river valley which favoured canal and railway expansion. Zone 5, high-quality housing, will extend along ridges of high ground or other pleasant environmental corridors. Whilst Hoyt's model is a step closer to reality, the **multiple nuclei model** is more so. No model will ever be an entirely successful simulation of real urban land-use patterns, as the local factors responsible for the structure of a given city will be unique to that location. *Compare* **shanty town**.

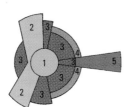

1 - Central business district

2 - Industrial zone

3 - Low quality housing

4 - Medium quality housing

5 - High quality housing

sector model

sediment   The material resulting from the **weathering** and **erosion** of the landscape, which has been deposited by water, ice or wind. It may be reconsolidated to form **sedimentary rock**.

sedimentary rock   A rock which has been formed by the consolidation of **sediment** derived from pre-existing rocks. **Sandstone** is a common example of a rock formed in this way; mudstone and shale are other examples. Such sedimentary rocks often show evidence of **bedding planes** which differentiate the layers of **deposition** sequences. **Chalk, limestone** and **evaporites** are other types of sedimentary rock, derived from organic and chemical precipitations.

seif dune   A linear sand dune, the ridge of sand lying parallel to the prevailing wind direction. The eddying movement of the wind keeps the sides of the dune steep. Seif dunes can be up to 200 m high and can run for over 10 km.

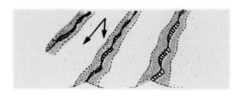

seif dunes

seismograph   An instrument that measures and records the seismic waves which travel through the Earth during an **earthquake**. The energy in these waves is measured on the **Richter scale**.

seismology   The study of **earthquakes**.

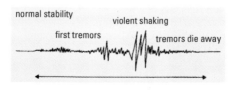

seismograph *A typical seismograph trace.*

serac   A pinnacle of ice formed by the tumbling and shearing of a **glacier** at an **ice fall**, i.e. the broken ice associated with a change in **gradient** of the valley floor.

service industry   The people and organizations that provide a service to the public. Examples are the transport industry, banking, the health service, the education service and the **retail industry**. The largest concentrations of service industries are found in urban areas. Service industries are in the **tertiary sector**.

settlement   Any location chosen by people as a permanent or semi-permanent dwelling place. Hence, settlements may vary from an individual farmhouse in an agricultural landscape, to a **conurbation** of several millions of people in an urban/industrial region.

settlement hierarchy   A series of size orders in a **settlement** system, the number of settlements in each order decreasing as the hierarchy is ascended. Generally, a settlement hierarchy will comprise many low-order places (or villages) and very few high-order places (or cities), with a number of intermediate orders between the two extremes. The logic behind the hierarchy concept is formalized in Walter Christaller's **central place theory**. Christaller suggested 'k' values to indicate the numerical relationship between one order and another; for example, in his k=3 hierarchy, the number of settlements is three times fewer at each successive higher order.

Such a hierarchy is, of course, entirely theoretical, but empirical studies do suggest some evidence of hierarchical organization.

Order	Number
4th order	1
3rd order	3
2nd order	9
1st order	27

settlement hierarchy

shading map or choropleth map A map in which shading of varying intensity is used. For example, the pattern of **population densities** in a region can be shown by means of a shading system.

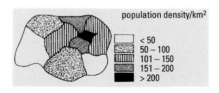

shading map

shaft mine A vertical mine. *Compare* **drift mine**.

shanty town   An area of unplanned, random, urban development often around the edge of a city. The shanty town is a major element of the structure of many **Third World** cities such as São Paulo, Mexico City, Nairobi, Calcutta and Lagos. The shanty town is characterized by high-density/low-quality dwellings, often constructed from the simplest materials such as scrap wood, corrugated iron and plastic sheeting – and by the lack of standard services such as sewerage and water supply, power supplies and refuse collection. The shanty town is the makeshift home of the relatively recent rural-urban migrant: unemployment tends to be high since the supply of urban employment is considerably less than the demand. However, the conventional image

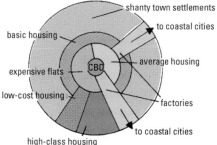

shanty town A model of a typical Third World city.

of the shanty town is misleading: the fact that such **settlements** survive is a measure of success; many occupants gradually improve their property as they become established in the urban sector, and the **informal economy** thrives.

A typical land-use pattern of a Third World city is given in the diagram on the left. *Compare* this with models for first-world cities such as the **Burgess model, sector model** and **multiple nuclei model**.

shifting cultivation   *See* **bush fallowing**.

shoreface terrace   A bank of **sediment** accumulating at the change of slope which marks the limit of a marine **wave-cut platform**.

Material removed from the retreating cliff base is transported by the **undertow** off the wave-cut platform to be deposited in deeper water offshore. (*See figure below*.)

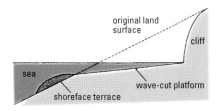

shoreface terrace

silage   Any **fodder crop** harvested whilst still green. The crop is kept succulent by partial fermentation in a *silo*. It is used as animal feed during the winter.

sill   **1.** An igneous intrusion of roughly horizontal disposition. Some sills form large, ridge-like escarpments when exposed; an example is the Great Whin Sill in northern England.
2. (Also called **threshold**) the lip of a **corrie**.

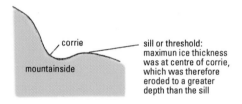

sill

silt   Fine **sediment**, the component particles of which have a mean diameter of between 0.002 mm and 0.02 mm.

sinkhole   *See* **pothole**.

slash and burn   *See* **tropical rainforest**.

slate   Metamorphosed shale or **clay**. Slate is a dense, fine-grained **rock** distinguished by the characteristic of *perfect cleavage*, i.e. it can be split along a perfectly smooth plane.

slip   The amount of vertical displacement of **strata** at a **fault**.

slip-off slope   The relatively gentle slope opposite a river cliff or **bluff** at a river **meander**. In the valley of a meandering river, the route of the fastest flowing water will undercut the outer bank of the meander. Eroded material from this bank may be deposited on the inner bank of the meander where the water is flowing at a slower speed. This deposited material forms a convex bank called a 'slip-off slope'.

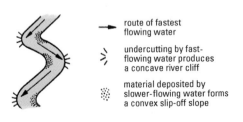

route of fastest flowing water

undercutting by fast-flowing water produces a concave river cliff

material deposited by slower-flowing water forms a convex slip-off slope

slip-off slope

smog   A mixture of smoke and fog associated with urban and industrial areas, that creates an unhealthy **atmosphere**. In 1952, a severe four-day smog in London left 4,000 people dead or dying. Since then, many cities introduced *clean air zones*. Los Angeles, in the USA, is renowned for its photochemical smog caused by the effect of the sun's heat on car exhaust emissions.

smokestack industry   A term used to describe any traditional heavy industry – iron and steel, shipbuilding – as opposed to newer, cleaner industries such as electronics.

snout   The end of a **glacier**; strictly the point at which wasting by melting equals the rate of supply of ice from up-valley. A glacier snout is characterized by a dissected, discoloured appearance caused by the action of meltwater and the washing out of **moraine**.

snow line   The altitude above which permanent snow exists, and below which any snow that falls will not persist during the summer months. The altitude of the snow line varies with **latitude**: at the equator it is approximately 5,000 metres, in the Alps it is approximately 3,000 metres, and at the Poles the snow line is at sea level.

socioeconomic group   A group defined by particular social and economic characteristics, such as educational qualifications, type of job, and earnings. In

census data, the socioeconomic groups are usually as follows: professional; employers and managers; non-manual; skilled workers; semi-skilled workers; and unskilled workers.

soil   The loose material which forms the uppermost layer of the Earth's surface, composed of the **inorganic fraction**, i.e. material derived from the **weathering** of bedrock, and the **organic fraction** – that is material derived from the decay of vegetable matter. The ultimate form of the organic fraction is humus. Soil also contains minerals and trace elements, and air and water held within the soil structure. *See* **horizon**. There are three broad categories of soil: *zonal*, i.e. the result of a specific combination of climatic and vegetative conditions; *intrazonal*, i.e. the result of local peculiarities of **environment** such as waterlogging or unusual parent material; and *azonal*, i.e. little-developed soils which occur on such surfaces as **scree, alluvium** or sand dunes.

soil erosion   The accelerated breakdown and removal of soil due to poor management. Soil erosion is particularly a problem in harsh **environments**. For example, in areas of steep slope, heavy seasonal rainfall, or strong dry-season winds, the soil may be rapidly eroded once protective vegetation has been removed. *See* **rainfall interception**. Severe soil erosion can cause problems of flooding and the silting-up of water supplies, especially in reservoirs. Erosion causes soil,

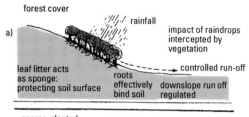

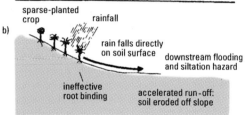

soil erosion *a) Stable environment, b) unstable environment.*

one of our most valuable **resources**, to be lost, much of it ending up in the sea.

Soil erosion by rainwater can be classified as follows: *splash erosion*, whereby the soil is pulverized by the impact of heavy raindrops and hailstones, as in a convectional storm; *sheet erosion*, whereby, in a heavy storm, a surface film of water develops which will flow downslope carrying surface soil as it moves; and *rill/gully erosion*, whereby any surface depression concentrates run-off, which quickly develops into channel flow, cutting a steep-sided valley as it runs off. Soil erosion by wind occurs on extensive flatlands which are subject to a windy, dry season for part of the year. The upper soil surface becomes loose and susceptible to wind erosion due to lack of moisture.

soil profile   The sequence of layers or **horizons** usually seen in an exposed soil section.

solar power   Heat radiation from the sun converted into electricity or used directly to provide heating. Solar power is an example of a renewable source of energy (*see* **renewable resources**).

solifluction   *See* **periglacial features**.

space   The geographer's term for area; the context within which distributions and patterns occur.

spatial analysis   The description and explanation of distributions of people and their activities in **space**.

spatial distribution   The pattern of locations of, for example, population or **settlement** in a region.

sphere of influence   The area, surrounding a **settlement**, from which consumers will travel in order to obtain goods and **services**. The size of the sphere of influence will vary according to the size of the settlement and the number and type of functions available there. For example, a neighbourhood shopping centre will have a small sphere of influence, i.e. people will not travel great distances to the centre as the range of goods and services will be limited. The central business district (**CBD**) of a town or city, however, will have a large sphere of influence. This means that people will travel from greater distances to the CBD, as the range of goods and services is much wider, including some specialized goods and services that cannot be obtained elsewhere.

spit   A low, narrow bank of sand and shingle built out into an **estuary** by the process of **longshore drift**. Spurn Head, Humberside, is a classic example. The **sediment** carried down the coast by longshore drift is deposited at the break in the coastline caused by the Humber estuary. Relatively shallow water, and the slowing of longshore drift by the counter-current of the Humber, result in the **deposition** of the marine **load**.

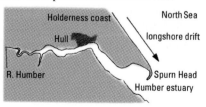

spit Spurn Head, a coastal spit.

The end of Spurn Head is 'hooked' by the action of waves swinging into the Humber estuary from the open North Sea.

spring   The emergence of an underground stream at the surface, often occurring where **impermeable rock** underlies **permeable rock** or **pervious rock** or **strata**.

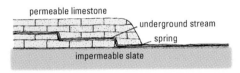

permeable limestone

underground stream

spring

impermeable slate

spring Rainwater enters through the fissures of the limestone and the stream springs out where the limestone meets slate.

spring line   A series of **springs** emerging along the foot of an **escarpment** of **permeable rock** or **pervious rock**. *Spring line settlements* can occur at these points, tapping the natural water supply. The settlements along the chalk escarpments of southern England are an example. (*See figure overleaf.*)

spring tides   *See* **tides**.

squatter settlement   An area of peripheral urban settlement in which the residents occupy land to which they have no legal title. *See* **shanty town**.

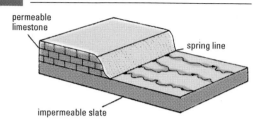

permeable limestone

spring line

impermeable slate

spring line

SSSI *abbreviation for* Site of Special Scientific Interest, an area containing rare species of plants or animals, or features of special geological interest.

SSSIs are designated in England by English Nature, but are mostly privately owned. English Nature has to be informed of developments being planned for SSSIs and can advise on their management. There are over 4,000 SSSIs in England, accounting for 7% of the total land area. Threats to SSSIs come from neglect, overgrazing and development.

stack A coastal feature resulting from the collapse of a **natural arch**. The stack remains after less resistant **strata** have been worn away by **weathering** and marine **erosion**. (*See figure on the right.*)

stalactite A column of calcium carbonate hanging from the roof of a **limestone** cavern. As water passes

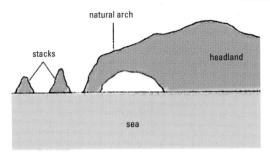

natural arch

stacks

headland

sea

stack

through the limestone it dissolves a certain proportion, which is then precipitated by **evaporation** of water droplets dripping from the cavern roof. The drops splashing on the floor of a cavern further evaporate to precipitate more calcium carbonate as a **stalagmite**.

stalagmite A column of calcium carbonate growing upwards from a cavern floor. *Compare* **stalactite**. Stalactites and stalagmites may meet, forming a column or pillar.

staple diet The basic foodstuff which comprises the daily meals of a given people. In South East Asia, rice is the staple food, while, in many parts of Africa the staple food is maize.

Stevenson's screen   A shelter used in weather
stations, in which thermometers and other instruments
may be hung. The screen consists of a wooden box
painted white and raised just over a metre from the
ground. The four sides are louvred to allow free entry of
air, and the roof is of double thickness to prevent the
sun's rays reaching inside.

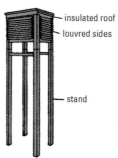

insulated roof

louvred sides

stand

Stevenson's screen

storm hydrograph   A graph showing the **discharge**
of a stream and its response to a period of rainfall.
There is a time lag between peak rainfall and peak
discharge. This is because it takes time for the rainwater
to pass through the various 'storage units', i.e.
vegetation, soil and rock. A series of storm hydrographs
can be plotted over several days to give a picture of the
river's *regime,* or pattern of flow.

strata   Layers of **rock** superimposed one upon the other; thus a sequence of strata is referred to as the *stratigraphy* of a region. The study of stratigraphy enables scientists to reconstruct the geological history of a region; this may be a complex process if the stratigraphy has been disturbed by earth movements such as folding or faulting.

stratosphere   The layer of the **atmosphere** which lies immediately above the troposphere and below the mesosphere and ionosphere. Within the stratosphere, temperature increases with altitude. The boundary between the stratosphere and the troposphere is known as the *tropopause*.

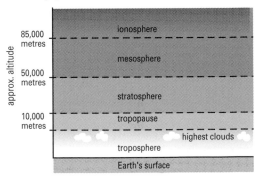

stratosphere

stratus   Layer-cloud of uniform grey appearance, often associated with the warm sector of a **depression**. Stratus is a type of low **cloud** which may hang as mist over mountain tops; broken stratus is referred to as *fractostratus*.

striations   The grooves and scratches left on bare **rock** surfaces by the passage of a **glacier**. Debris embedded in the glacier scores the surface over which it passes. Striations may thus be a guide to the direction of ice movement.

strip cropping   A method of **soil** conservation whereby different crops are planted in a series of strips, often following **contours** around a hillside. The purpose of such a sequence of cultivation is to arrest the downslope movement of soil, especially if one of the crops (e.g. maize) tends to expose the soil surface. Alternate strips may be planted with grass which effectively binds the soil and acts as a brake on run-off. *See* **soil erosion**.

subduction zone   *See* **plate tectonics**.

subsidiary cone   A volcanic cone which develops within the **caldera** of a previous, larger cone which was blown away during an eruption.

subsistence agriculture   A system of **agriculture** in which farmers produce exclusively for their own consumption, in contrast to **commercial agriculture** where farmers produce purely for sale at the market.

subsoil   *See* **soil profile**.

suburbs   The outer, and largest, parts of a town or city. They consist mainly of low-density housing (either privately owned houses or council houses owned by the local authority). As the suburbs have spread outwards into the surrounding countryside, they have sometimes engulfed formerly rural villages. Suburbs also contain shopping centres, open spaces, schools, leisure facilities, etc. Offices, warehouses and small firms may also locate in the suburbs, taking advantage of the lower land values compared with the **CBD**. Beyond the suburbs, many towns and cities have a **green belt** to stop **urban sprawl**.

sunshine recorder   An instrument used to measure the hours of sunshine in an area. It consists of a glass sphere mounted on a frame. The sun's rays are concentrated by the sphere on to a strip of sensitized paper behind. This paper is graduated into hours. When the sun shines, a line is burnt along the paper as the Earth rotates. The total length of the burnt line will thus indicate how many hours in the day the sun has shone. On a meteorological map, parallel lines drawn through places having the same amount of sunshine are called *isohels*.

surface run-off That proportion of rainfall received at the Earth's surface which runs off either as channel flow or overland flow. It is distinguished from the rest of the rainfall, which either percolates into the soil or evaporates back into the **atmosphere**.

sustainable development The ability of a country to maintain a level of economic development, thus enabling the majority of the population to have a reasonable standard of living.

swallow hole *See* **pothole**.

swash The rush of water up the beach as a wave breaks. See also **backwash** and **longshore drift**.

syncline A trough in folded **strata**; the opposite of **anticline**. *See* **fold**.

T

taiga The extensive **coniferous forests** of Siberia and Canada, lying immediately south of the arctic **tundra**. Within the taiga there are many lakes, marshes and swamps, the latter often resulting from springtime thaw occurring over permafrost and while river courses to the north are still frozen. *See also* **periglacial features**.

tarn The postglacial lake which often occupies a **corrie**. *See figure below.*

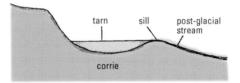

temperate climate A climate typical of mid-latitudes. The British Isles, for example, have a temperate climate. Such a climate is intermediate between the extremes of hot (tropical) and cold (polar) climates. Compare **extreme climate**. *See also* **maritime climate**.

terminal moraine *See* **moraine**.

terrace 1. *See* **river terrace**.

2. A type of housing in which all the dwelling units are joined together at each side, as opposed to detached or semi-detached housing.

terracing   A means of **soil** conservation and land utilization whereby steep hillsides are engineered into a series of flat ledges which can be used for **agriculture**, held in places by stone banks to prevent **soil erosion**. Terracing reaches its most sophisticated and intricate form in the fertile volcanic hills of Java, Indonesia, where a subsistence rice economy prevails.

terracing

tertiary sector   That sector of the economy which provides **services** such as transport, finance and retailing, as opposed to the **primary sector** which provides **raw materials**, the **secondary sector** which processes and manufactures products, and the **quaternary sector** which provides information and expertise. In highly developed economies the tertiary sector is the dominant employer, but in recent years numbers employed in the quaternary sector have gone up due to use of computers and electronic information.

textile industry   The manufacture of cloth and related materials. Traditionally, the textile industry used wool, cotton, flax and other natural products as its **raw materials**, but this sector has declined with the rise in use of **artificial fibres** derived from oil and other hydrocarbons. Nylon, rayon and polyester are examples of such fibres

thermal power station   An electricity-generating plant which burns **coal**, oil or natural gas to produce steam to drive turbines.

Third World   A collective term for the poor nations of Africa, Asia and Latin America, as opposed to the 'first world' of capitalist, developed nations and the 'second world' of formerly communist, developed nations. The terminology is far from satisfactory as there are great social and political variations within the 'Third World'. Indeed, there are some countries where such extreme poverty prevails that these could be regarded as a fourth group. Alternative terminology includes '**developing countries**', 'economically developing countries' and 'less economically developed countries' (LEDC). **Newly industrialized countries** are those showing greatest economic development.

threshold   *See* **sill** (sense 2).

threshold population   The minimum population required in a **sphere of influence** to sustain a particular good or service offered in a **central place**. The size of the threshold population increases with progressively more specialized goods and services for which individual demand is less frequent.

tidal limit   The point in a river upstream of which there is no tidal rise and fall in river level.

tidal range   The mean difference in water level between high and low tides at a given location. *See* **tides**.

tides   The alternate rise and fall of the surface of the sea, approximately twice a day, caused by the gravitational pull of the moon and, to a lesser extent, of the sun.

When the moon, earth and sun are in alignment, at full and new moon, the gravitational force is at its greatest because the pull of the sun is combined with

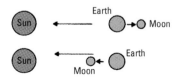

tides *Spring tide.*

***tides** Neap tide.*

that of the moon. This produces a *spring tide*, characterized by a large tidal range (very high tides and equally low tides). When the moon, earth and sun are not aligned, the effect on the Earth is less. The pull of the sun and of the moon are at right angles when there is a half moon. At this time, the difference between high and low tides is not very great. These tides are called *neap tides*.

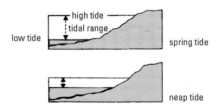

***tides** Tidal ranges.*

till *See* **boulder clay**.

tombolo   A **spit** which extends to join an island to the mainland, as in the case of Chesil Beach, Portland Island, southern England.

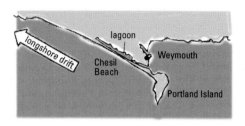

tombolo Chesil Beach.

topography   The composition of the visible landscape, comprising both physical features (e.g. relief, **drainage** and vegetation) and those made by people (e.g. roads, railways and **settlements**).

topsoil   The uppermost layer of **soil**, more rich in organic matter than the underlying **subsoil**. *See* **horizon**, **soil profile**.

tourist industry   The people and organizations that provide facilities and services for tourists.
Tourism is a major industry in many countries of the world, employing millions of people, either directly or indirectly.

In some developing countries, the tourist industry has brought in capital, but tourism on a large scale can have detrimental effects on the traditional culture and lifestyle of the indigenous population, and extensive hotel-building damages the countryside. Some countries, such as Tanzania, have developed a form of eco-tourism designed to encourage specialized visitors, rather than mass-market tourism, as a way of protecting the environment. Large numbers of tourists to any area can cause problems, ranging from footpath erosion in a **national park** such as the Lake District, to congested roads and air **pollution** by motor vehicles. Tourism and **conservation** can come into conflict over the destruction either of the natural environment or of a city such as Venice.

trade winds   Winds which blow from the subtropical belts of high pressure towards the equatorial belt of low pressure. In the northern hemisphere, the winds blow from the northeast and in the southern hemisphere from the southeast. In most areas, these winds blow with great regularity, and weather in the trade wind regions is usually fine and quiet. However, these regions can sometimes experience **hurricanes**.

transhumance   The practice whereby herds of farm animals are moved between regions of different climates. Pastoral farmers (*see* **pastoral farming**) take their herds from valley pastures in the winter to mountain pastures in the summer. Frequently, farmers will live in mountain huts during the summer in mountainous regions such as the Himalayas and the Alps.

transnational corporation (TNC)   A company that has branches in many countries of the world, and often controls the production of the primary product and the sale of the finished article. For example, TNCs own many tea plantations in the Third World, where the tea is picked and processed by local labour (*see* **plantation agriculture**). The TNCs also control the selling of the tea, most of which is sent to developed countries. It is argued by some economists that labour in the Third World is badly paid in order to provide cheap products worldwide. The TNCs, however, claim to provide higher wages and better working conditions than indigenous industries.

transpiration   The process whereby plants give off water vapour via the stomata of their leaves. Water taken up by roots is thus returned to the **atmosphere**.

tropic   *See* **latitude**.

tropics   The region of the Earth lying between the *tropics of Cancer* (23° 27'N) and *Capricorn* (23° 27'S). *See* **latitude**.

tropical rainforest   The dense forest cover of the equatorial regions, reaching its greatest extent in the Amazon Basin of South America, the Congo Basin of

Africa, and in parts of South East Asia and Indonesia. The lush forest is a response to optimum climatic conditions (high temperatures and abundant moisture), and not to soil fertility. Tropical soils are, in fact, generally poor. Newly germinated seedlings grow rapidly upwards in search of light in the dense forest cover, before branching out having reached the forest canopy. Most trees are very shallow rooted, and are often buttressed to provide support. The richness and diversity of the forest itself gives rise to a similarly varied fauna. There has been much concern in recent years about the rate at which the world's rainforests are

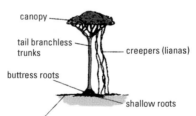

canopy

tail branchless trunks

creepers (lianas)

buttress roots

shallow roots

Intense bacterial activity breaks down fallen leaves, etc., to return nutrients to soil surface for immediate uptake by roots. Soils themselves are infertile: the nutrient cycle is concentrated in the vegetation and top few inches of soil.

a forest giant in the tropical rainforest

being cut down. There are several reasons for this
deforestation, such as *slash and burn* farming by the local
population, and, more especially, large-scale forest
clearance to provide land for cattle farming or for
development projects such as **hydroelectric power**
schemes and mineral excavation. The burning of large
tracts of rainforest is thought to be contributing to
global warming. Many governments and **conservation**
bodies are now examining ways of protecting the
remaining rainforests, which are unique **ecosystems**
containing millions of plant and animal species.

troposphere   *See* **atmosphere**.

trough   An area of low pressure, not sufficiently well-
defined to be regarded as a **depression**.

trough end wall   The steep rear wall of a **U-shaped
valley**, formed where coalescing **corrie** glaciers cause an
increase in **erosion** and a consequent deepening of the
glacial valley.

truncated spur   A spur of land that previously
projected into a valley and has been completely or
partially cut off by a moving **glacier**. The removal of
spurs in a river valley therefore has the effect of
widening, straightening and deepening the valley. *See*
U-shaped valley. The position of the truncated spur is
usually marked by a **crag** and **scree**.

tsunami   A very large, and often destructive, sea wave produced by a submarine **earthquake**. Tsunamis tend to occur along the coasts of Japan and parts of the Pacific Ocean, and can be the cause of large numbers of deaths. Tsunamis are sometimes incorrectly referred to as 'tidal waves'.

tuff   Volcanic ash or dust which has been consolidated into **rock**.

tundra   The barren, often bare-rock plains of the far north of North America and Eurasia where subarctic conditions prevail and where, as a result, vegetation is restricted to low-growing, hardy shrubs and mosses and lichens. Permafrost conditions result in poor **drainage** which results in marsh and swamp during the short summer. *See* **periglacial features**.

U

undernutrition A lack of a sufficient quantity of food, as distinct from **malnutrition** which is a consequence of an unbalanced diet. Many of the world's poorest people suffer both undernutrition and malnutrition. As the world's population rises at a faster rate than that of the food supply, problems of undernutrition are becoming worse in many parts of Latin America, Africa and Asia.

undertow The counter-current to water breaking onshore as waves. The undertow is responsible for the removal of eroded material from the **wave-cut platform**, to be deposited as the **shoreface terrace**. The undertow operates at a much larger scale than **backwash**.

urban decay The process of deterioration in the **infrastructure** of parts of the city – especially in the old industrial cities of, for example, Northern England and the Midlands. It is the result of long-term shifts in patterns of economic activity, residential **location** and **infrastructure**.

Parts of the inner city are especially decayed: old Victorian **terrace** housing, mills and other traditional industrial installations such as disused canals and warehouses.

Much of the decay has resulted from neglect, as the focus of the urban system has moved away from these areas – for example, towards new peripheral locations for industry, newer housing in the outer **suburbs**, and new **communications** bypassing the city. In order to reverse the problem, many cities have undertaken 'face-lift' schemes to improve the decayed **environments**, either by demolition and landscaping, or by renovating old property for new uses. *See* **comprehensive redevelopment**.

urban sprawl   The growth in extent of an urban area in response to improvements in transport and rising incomes, both of which allow a greater physical separation of home and work. The sprawling outer suburbs of cities in developed countries today are a result of almost universal car ownership; in earlier years similar, though smaller, expansions of the urban area resulted from the development of the urban railway, the tram and the bus. (*See figure overleaf.*) Urban sprawl has caused towns to merge. *See* **conurbation**, *also* **suburb** and **green belt**.

U-shaped valley   A glaciated valley, characteristically straight in plan and U-shaped in **cross section**.

The winding V-shaped valley of a river's course is modified into a U-shaped valley by the greater erosive power of a **glacier**.

late
⑲

low mobility,
majority of urban
population walked
to work

mid
⑳

spreading city
in response to
developing
public transport

2000

high mobility
due to higher incomes
and car ownership

urban sprawl

urbanization   The process by which a national
population becomes predominantly urban through a
migration of people from the countryside to cities,
and a shift from agricultural to industrial employment.
Urbanization is thus an important element of the
development process. The process of **gentrification** has
also led to the restoration of housing in some inner city
areas. *Contrast* **counter-urbanization**.

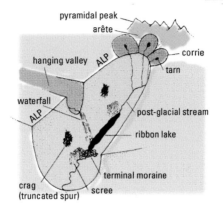

U-shaped valley

V

vicious cycle of poverty The poverty trap in
which much of the population of the **Third World** finds
itself. Poor farmers cannot afford to invest in their land
through improved seed or fertilizer: **yields** thus remain
low, there is no surplus for sale at market, and so the
poverty continues. In the absence of credit or grant aid,
there is no way in which the farmer can break out of
the cycle.

Given that poor yields also lead to a food shortage,
the farmer will suffer poor nutrition and may even

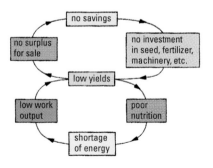

vicious cycle of poverty

have to borrow money to secure enough food for
subsistence, let alone invest in the land. It is not only
necessary to provide the financial means for the farmer
to improve his or her livelihood, it is also important to
provide the correct institutional context for progress –
for example, **land reform** may be necessary to ensure
security of tenure.

viscous lava   Lava that resists the tendency to
flow. It is sticky, flows slowly and congeals rapidly. *Non-
viscous* lava is very fluid, flows quickly and congeals
slowly.

volcanic rock   A category of **igneous rock** which
comprises those rocks formed from **magma** which has
reached the Earth's surface. (This is to be contrasted
with **plutonic rock** which forms below the surface.)
Basalt is an example of a volcanic rock, as are all
solidified lavas.

volcano   A fissure in the Earth's **crust** through which
magma reaches the Earth's surface. There are four main
types of volcano:
(a) *Acid lava cone* – a very steep-sided cone composed
entirely of acidic, **viscous lava** which flows slowly and
congeals very quickly. The cones of the Puy region in
central France are classic examples.
(b) *Composite volcano* – a single cone comprising
alternate layers of ash (or other pyroclasts) and lava.
Such volcanos as Vesuvius and Etna are of this type.

volcano *Composite volcano.*

c) *Fissure volcano* – a volcano that erupts along
a linear fracture in the crust, rather than from a single
cone. Fissure volcanos are generally of a quiet,
unexplosive nature, relating to the non-viscous lavas
which are usually produced. Many eruptions in Iceland
are of this type.
(d) *Shield volcano* – a volcano composed of very basic,
non-viscous lava which flows quickly and congeals
slowly, producing a very gently sloping cone. Mauna
Loa, Hawaii, is a good example of this type.

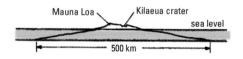

volcano *Shield volcano.*

Von Thünen theory J.H. Von Thünen's early-19th-century model of the distribution of agricultural land use around a town is still the basis for many investigations of such land-use patterns. Von Thünen envisaged a single market town, surrounded by a region supplying agricultural produce. Physical and economic characteristics of this agricultural area were regarded as uniform.

1 - market gardening
2 - dairying
3 - arable
4 - rough grazing
5 - wilderness

Von Thünen theory *Agricultural land use around a market town.*

Such conditions of course do not exist in reality, but the value of a model like Von Thünen's is that it sweeps away the clutter of reality and allows us to observe basic processes at work. Von Thünen postulated the following land-use system, given his simplifying assumptions, illustrated in the diagram.

The zones 1-5 are derived from bid-rent curves thus:

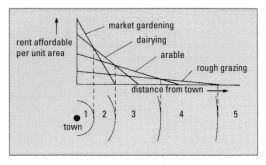

Von Thünen theory Bid-rent curves.

The implication of the theory is that the distribution of land uses around the town will depend upon a variety of factors: perishability, transport costs and land-area requirements.

Highly perishable produce such as salad crops must be grown close to town to ensure freshness at market and to minimize the transport costs necessitated by frequent marketing. The location of market gardens close to the town is also explained by the fact that land-area requirements are small, and therefore the bid-rent per unit area can be high. Market gardeners can achieve high productivity from small areas by the intensive nature of their farming, and through investment in greenhouses, fertilizers and other inputs. Land uses with progressively larger land-area requirements must locate further away from the town; a farmer needing many units of land will only be able to bid a lower rent per unit area. Land uses at progressively greater distance from town are also characterized by a decreasing frequency of marketing of products, and by decreasing in the investment land. In reality, such patterns are constrained by variations in the physical **environment** and in human behaviour; government policy also affects farming practices and the distribution of land uses.

V-shaped valley   A narrow, steep-sided valley made by the rapid erosion of rock by streams and rivers. It is V-shaped in cross-section and is not a glacial feature. *Compare* **U-shaped valley**.

vulcanicity   A collective term for those processes which involve the intrusion of **magma** into the **crust**, or the extrusion of such molten material onto the Earth's surface.

W

wadi A dry watercourse in an arid region; occasional rainstorms in the desert may cause a temporary stream to appear in a wadi.

warm front *See* **depression.**

warm sector *See* **depression.**

waterfall An irregularity in the long profile of a **river's course**, usually located in the upper course. A waterfall occurs either at the edge of a stratum of resistant rock (**cap rock**), where it forms a lip, the softer rock immediately downstream being eroded more quickly than the resistant rock; or where strata of resistant rock **dip** upstream (*compare* **rapid**): again the softer rock downstream wears away sooner, and a step is formed. (*See figure below.*)

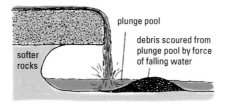

plunge pool

debris scoured from plunge pool by force of falling water

softer rocks

water gap   In chalk **escarpments**, a valley which has been eroded below the depth of the **water table**, and which therefore contains a permanent stream. *Compare* **dry valley**, **wind gap**.

watershed   The boundary, often a ridge of high ground, between two **river basins**.

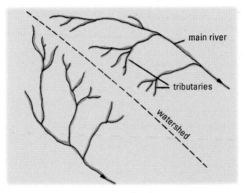

watershed

water table   The level below which the ground is permanently saturated. The water table is thus the upper level of the **groundwater**. In areas where **permeable rock** predominates, the water table may be at some considerable depth. In periods of high rainfall the level of the water table will rise; it will fall in protracted dry spells.

wave refraction   The bending of waves around a **headland**. The shorewards movement of water in contact with the headlands is slowed, while open water in the middle of the bay moves on unimpeded.

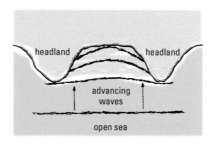

wave refraction

wave-cut platform or abrasion platform

A gently sloping surface eroded by the sea along a coastline. As the **cliff** recedes, the wave-cut platform is created. The platform slopes seawards, since **erosion** has been active for a longer period of time at the seaward end. When sea levels are lowered, a wave-cut platform may appear as a *raised beach* several metres above the current sea level. Raised beaches can be seen on the western coast of the Kintyre peninsula in Scotland.

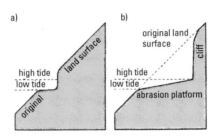

wave-cut platform a) Early in formation, b) later in formation.

weather The day-to-day conditions of e.g. rainfall, temperature and pressure, as experienced at a particular location. Contrast this with **climate**, which is a set of long-term, average atmospheric conditions. Thus the climate of, for example, a location in the Sahara **desert** can be characterized as hot and dry all the year round, but there will be occasions when violent weather in the form of convectional rainstorms occurs.

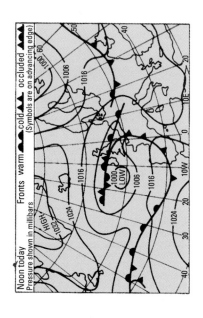

weather chart

weather chart   A map or chart of an area giving details of **weather** experienced at a particular time of day, such as 0600 hrs. The information is gathered at weather stations throughout the country, and when all observations have been collected **isobars** can be drawn and the **depressions, anticyclones, fronts**, etc. can be identified. Weather charts are sometimes called *synoptic charts*, as they give a synopsis of the weather at a particular time.

weather station   A place where all elements of the weather are measured and recorded. Each station will have a **Stevenson's screen** and a variety of instruments such as a **maximum and minimum thermometer**, a **hygrometer**, a **rain gauge**, a **wind vane**, an **anemometer** and a **sunshine recorder**.

weathering   The breakdown of rocks *in situ*; contrasted with **erosion** in that no large-scale transport of the denuded material is involved. Weathering processes include **exfoliation, nivation** and chemical activity such as the dissolving of **limestone** by rainwater. Biological activity, for example by tree roots and earthworms, also contributes towards the breakdown of bedrock. Thus, three types of weathering are identified: mechanical (or physical), chemical and biological.

Weber's theory   *See* **industrial location**.

wet and dry bulb thermometer   *See* **hygrometer**.

white ice   Ice from which air has not been totally expelled (*see* **corrie**). Contrast this with **blue ice** which is found at greater depth in a corrie icefield and from which air has been expelled by compression.

white-collar worker   A worker who is not a manual worker and who does not work in 'dirty conditions'. The term 'white-collar' derives from the white shirts worn by e.g. clerical workers and professional people whose jobs do not make their clothes dirty. *Compare* **blue-collar worker**.

wind gap   A **dry valley**, for example in a chalk **escarpment**, now standing above the **water table**, but formed at a time when the water table was higher or when the ground was frozen.

wind vane   An instrument used to indicate wind direction. It consists of a rotating arm which always points in the direction from which the wind blows. The wind is named after this direction.

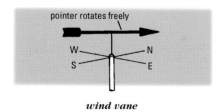

wind vane

Y

yardang   Long, roughly parallel ridges of **rock** in arid and semi-arid regions. The ridges are undercut by wind **erosion** and the corridors between them are swept clear of sand by the wind. The ridges are oriented in the direction of the prevailing wind. An area containing yardangs is often referred to as a 'ridge and furrow' landscape. The term 'yardang' comes from Central Asia.

yield   The productivity of land as measured by the weight or volume of produce per unit area. Agricultural yields are usually expressed per hectare.

Z

Zeugen   *Pedestal rocks* in arid regions; wind **erosion** is concentrated near the ground, where **corrasion** by wind-borne sand is most active. This leads to undercutting and the pedestal profile emerges.

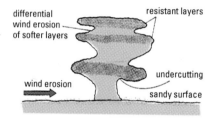

Zeugen